The Role of Standards in Fostering Capability Evolution

Does Design Matter? Insights from Interoperability Standards

JON SCHMID, BONNIE L. TRIEZENBERG, JAMES DIMAROGONAS, SAMUEL ABSHER

Prepared for the Office of the Under Secretary of Defense for Research and Engineering
Approved for public release; distribution unlimited

NATIONAL DEFENSE RESEARCH INSTITUTE

For more information on this publication, visit **www.rand.org/t/RRA1576-1**.

About RAND

The RAND Corporation is a research organization that develops solutions to public policy challenges to help make communities throughout the world safer and more secure, healthier and more prosperous. RAND is nonprofit, nonpartisan, and committed to the public interest. To learn more about RAND, visit www.rand.org.

Research Integrity

Our mission to help improve policy and decisionmaking through research and analysis is enabled through our core values of quality and objectivity and our unwavering commitment to the highest level of integrity and ethical behavior. To help ensure our research and analysis are rigorous, objective, and nonpartisan, we subject our research publications to a robust and exacting quality-assurance process; avoid both the appearance and reality of financial and other conflicts of interest through staff training, project screening, and a policy of mandatory disclosure; and pursue transparency in our research engagements through our commitment to the open publication of our research findings and recommendations, disclosure of the source of funding of published research, and policies to ensure intellectual independence. For more information, visit www.rand.org/about/principles.

RAND's publications do not necessarily reflect the opinions of its research clients and sponsors.

Published by the RAND Corporation, Santa Monica, Calif.
© 2022 RAND Corporation
RAND® is a registered trademark.

Library of Congress Cataloging-in-Publication Data is available for this publication.

ISBN: 978-1-9774-0878-5

Cover: Alex/Adobe Stock and USAF photo.

About This Report

The U.S. military is currently engaged in the design of new command and control interface standards in the belief that *joint all-domain* or *universal standards* will improve interoperability between existing and future systems and unleash a new wave of evolution in how warfighters interact to defend U.S. interests and defeat adversaries. The role that interface standards play in fostering interoperability and innovation has been the subject of considerable study. It is generally accepted that today's internet protocol standards have enabled one of the greatest technological and social revolutions in history, fundamentally changing how humans work, learn, and interact. Yet, history also offers examples in which standards hindered innovation by enshrining the status quo—simply standardizing an interface is not enough to enable interoperability and innovation. Therefore, to tease out potential explanations for which standards enable interoperability and innovation and which do not, in this report we address a fundamental question: What features of interface standards foster interoperability and capability evolution? This report should be of interest to all engaged in the development, design, and deployment of interoperability standards within the U.S. military and defense industry.

The research reported here was completed in October 2021 and underwent security review with the sponsor and the Defense Office of Prepublication and Security Review before public release.

RAND National Security Research Division

This research was sponsored by the Office of the Under Secretary of Defense for Research and Engineering and conducted within the Acquisition and Technology Policy Center of the RAND National Security Research Division (NSRD), which operates the National Defense Research Institute (NDRI), a federally funded research and development center sponsored by the Office of the Secretary of Defense, the Joint Staff, the Unified Combatant Commands, the Navy, the Marine Corps, the defense agencies, and the defense intelligence enterprise.

For more information on the RAND Acquisition and Technology Policy Center, see www.rand.org/nsrd/atp or contact the director (contact information is provided on the webpage).

Acknowledgments

We thank Michael Zatman, assistant director, Fully Networked Command, Control and Communications, Office of the Under Secretary of Defense for Research and Engineering, for entrusting us with this study. Our study benefited from conversations with various technical experts, including Brian R. O'Hern, Steven Bygren, Chris Ramirez, Brian E. White, Carl Nordstrom, Nicholas A. O'Donoughue, and Randy Burnham. For their contributions in editing and formatting the document, we thank Rosa Maria Torres. Finally, we thank the Acquisition and Technology Policy Center leadership team, Joel Predd, Caitlin Lee, and Yun Kung, for their support. Although we acknowledge the contributions of all the above, we remain solely responsible for the quality, integrity, and objectivity of our assessment and recommendations.

Summary

The U.S. military is engaged in the design of new command and control interface standards in the belief that *joint all-domain* or *universal standards* will improve interoperability between existing and future systems and unleash a new wave of evolution in how warfighters interact to defend U.S. interests and defeat adversaries. The role that interface standards play in fostering interoperability and innovation is a subject of considerable study. It is generally accepted that today's internet protocol standards have enabled one of the greatest technological and social revolutions in history, fundamentally changing how humans work, learn, and interact. Yet, history also offers examples in which standards have hindered innovation by enshrining the status quo—simply standardizing an interface is not enough to enable interoperability and innovation. Therefore, to tease out potential explanations for which standards enable interoperability and innovation, we address a fundamental question: What features of interface standards foster interoperability and capability evolution? This report should be of interest to all engaged in the development, design, and deployment of interoperability standards within the U.S. military and defense industry.

Scholarship regarding the role of interface standardization is largely focused on the use of standards in commercial for-profit ecosystems, with little research on how military standards affect interoperability and capability evolution. Given that military actors are embedded in a largely nonmarket context, separate consideration of military standards is warranted. In this report, we examine two military standards—(1) Standardization Agreement (STANAG) 4607, which establishes North Atlantic Treaty Organization (NATO)'s ground moving target indicator (GMTI) data format, and (2) the Link 16 messaging standard—that have enabled significant interoperability and evolution in warfighting. Our goal is to understand a range of paths that could lead to successful standard design in the military environment. We found that, despite significant variation in design paradigms, there are common characteristics in the design of the standards and governance processes that were critical to the evolutions that these characteristics enabled. These characteristics are as follows:

- **built-in extensibility of the standard.** Built-in extensibility allowed the standards to later include data from technologies that had not yet matured when the standard was originally designed. For example, new message segments were added to the STANAG 4607 standard to accommodate data from space-based radars.
- **formal, transparent, and relatively open processes for extending and amending the standard.** The standards have clear and open processes for amendments or additions. For example, the STANAG 4607 standard documents describe how to include new capabilities, such as radar modes, platform types, or data processing techniques. This process is open to any national representative and simply requires formally proposing a change to the STANAG 4607 custodial support team.
- **the use of operational feedback in the design and amendment of the standard.** Feedback from real-world operations drove evolution of the standards. For example, platform- and service-specific implementations of the Link 16 message standard led to communication problems during the 1990–1991 Gulf War. Because of this experience, ambiguity in the Link 16 message standard implementation guidance was removed.
- **access to standardized data that allows a community of active and engaged users and suppliers to evolve capabilities rapidly.** For example, the STANAG 4607 user community, aided by the Atlantis GMTI data repository, developed several applications for the new operational domain of counter-improvised explosive devices. These applications included algorithms for group tracking, forensic backtracking, meeting detection, route avoidance, anomaly detection, and road and traffic detections.

Our suggested guidance on standard design and governance follows directly from these findings. To enable interoperability and capability evolution, we suggest that standards and their governing bodies

- be designed with a built-in technical means of extensibility
- lay out an open and transparent means of drafting and amending the standard
- seek feedback from real-world operations at all stages of the standard's life cycle
- cultivate a user and supplier community with the means of providing feedback into the standard design and amendment process.

Regarding design paradigms of the standards themselves, we found that standardization of data yields innovation, while standardization of transport and link layers has inhibited innovation. We believe this is because data are a resource that can be exploited, while transports and links are constraints that must be overcome. Therefore, our design guidance mirrors two of the data decrees from a U.S. Department of Defense official's May 2021 memorandum about creating data advantage:

- Treat data as assets, which include the creation of common interface specifications.
- At the transport and link layers, "use automated data interfaces that are externally accessible and machine-readable; [. . .] use industry-standard, non-proprietary, preferably open-source technologies, protocols, and payloads."[1]

---

[1] Kathleen Hicks, "Creating Data Advantage," memorandum from the Deputy Secretary of Defense to Senior Pentagon Leadership Commanders of the Combatant Commands Defense Agency and U.S. Department of Defense Field Activity Directors, Washington, D.C., May 5, 2021.

Contents

Figures and Tables

Figures

Tables

CHAPTER ONE

Introduction

Research Goal

The 2018 National Defense Strategy lists command, control, communications, computers and intelligence, surveillance, and reconnaissance (C4ISR) as one of the U.S. Department of Defense (DoD)'s principal modernization priorities.[1] To this end, the Office of Under Secretary of Defense for Research and Engineering (OUSD-R&E) has been charged by DoD with developing a long-term (beyond 2024) fully networked command, control, and communications (FNC3) architecture to underlie the C4ISR capability.[2] FNC3 is composed of three dimensions: universal command and control (UC2), fully networked link diversity and software-defined networking, and multifunctional radio frequency and optical systems (e.g., systems that can simultaneously provide sensing, command and control, and electronic warfare capability).[3] This report seeks to inform the UC2 aspect of the FNC3 architecture and DoD standards development more generally.

The UC2 standard is an application-layer message standard for machine-to-machine command and control. The proposed UC2 standard intends to replace older stovepiped interfaces between weapons systems with a new language focused on interoperability. The UC2 will be designed to support future innovation and evolution as DoD moves beyond individually acquired systems supporting a single mission or service branch into more-composable system of systems (SoS) that support the joint force. DoD believes improved interoperability between existing and future systems will unleash a new wave of evolution in how warfighters interact to defend U.S. interests and to deter and defeat adversaries.

Taking a lesson from the way the internet protocol standards and layering have unleashed evolution in other domains—such as banking, retail, and communication—the FNC3 standards are designed to ensure that network, transport, link, and application layers remain open and separate. The UC2 element of the FNC3 is explicit in its intention to enable warfighting evolution and innovation at the application layer independent of other layers of the computing infrastructure, as shown in Figure 1.1.

Although the example of the internet is one illustration of a standard fostering innovation, history also offers examples in which standards have hindered evolution by enshrining the status quo—i.e., simply standardizing an interface is not enough to enable interoperability and capability evolution.[4] Therefore, in this

---

[1] James Mattis, *Summary of the 2018 National Defense Strategy of the United States of America*, Washington, D.C.: U.S. Department of Defense, 2018.

[2] U.S. Department of Defense, *C3 Command, Control, and Communications: Modernization Strategy*, Washington, D.C., September 2020, p. 3.

[3] Michael D. Griffin, "Statement of Michael D. Griffin, Under Secretary of Defense for Research and Engineering," statement before the House Armed Services Subcommittee on Intelligence, Emerging Threats and Capabilities, FY2020 Science and Technology Posture Hearing, Washington, D.C., March 11, 2020.

[4] One example of a standard appearing to hinder innovation is the QWERTY keyboard layout. This example is described in greater detail in Chapter Two.

FIGURE 1.1

Network-Layering Abstractions and Terms

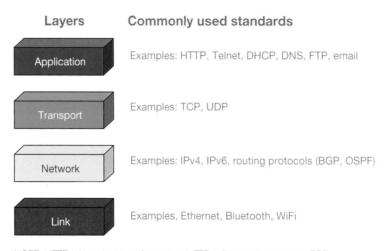

NOTE: HTTP = hypertext transfer protocol; FTP = file transfer protocol; TCP = transmission control protocol; UDP = user datagram protocol; IP = internet protocol; IPv4 = internet protocol version 4; IPv6 = internet protocol version 6; BGP = border gateway protocol; OSPF = Open Shortest Path First; WiFi = wireless fidelity.

report, we pose a research question that aims to clarify what variables explain this variation in outcomes: What features of interface standards foster interoperability and capability evolution?

Although the independence of the application, transport, network, and link layers may be one key to enabling innovation and evolution of systems, other factors may be just as influential in enabling the innovation and evolution of warfighting. Other attributes, such as simplicity, extensibility, modularity, or maintainability, may be just as determinant of whether a standard enables the innovation envisioned by DoD.

Prior scholarship on this topic has focused on the use of standards in commercial for-profit ecosystems, but it is unclear whether the attributes of standards that enable success in those environments apply fully to military systems where actors have different motivations. Therefore, in this report, we examine two military standards (both of which have successfully enabled significant interoperability and evolution in warfighting) to understand a range of paths that could lead to success in the military environment. We anticipate that this work will illuminate future OUSD-R&E design decisions regarding the implementation of the UC2 interfaces or decisions regarding the design and governance of weapons systems interfaces more generally.

Standards and Warfighting Evolution

This study does not seek to broadly assess, or quantify, the determinants of military innovation. Myriad factors besides the design of interfaces influence whether systems evolve and whether an SoS grows up around that system. Within the context of military technology, factors found to drive warfighting evolution include changes in military doctrine, intraservice branch competition, interservice branch competition, a strong high-ranking champion, external threats, and a compelling operational need or requirement.[5] To our knowledge, the potential effect of interface design on military innovation has not yet been investigated.

---

5 Dennis J. Blasko, "'Technology Determines Tactics': The Relationship Between Technology and Doctrine in Chinese Military Thinking," *Journal of Strategic Studies*, Vol. 34, No. 3, 2011; Stephen P. Rosen, *Winning the Next War: Innovation and the Modern Military*, Ithaca, New York: Cornell University Press, 1994, Harvey M. Sapolsky, *The Polaris System Development:*

An investigation of how military standards affect military warfighting evolution is motivated, in part, by the observed relationship between standards and commercial innovation.[6] As noted earlier in this report, although independence of application, network, and link layers may be one key to enabling innovation and evolution of systems, other factors may be at least as influential in enabling the innovation and evolution of warfighting. In particular, the Link 16 message standard we selected as one of our case studies appears to have succeeded in transforming warfighting only after removing some of the interdependencies between these layers.

Case Study Selection

This research is exploratory and does not seek to make inferential or causal claims. Instead, we use a case study approach to understand the variety of approaches used in the design of both standards and standard governance that might affect later capability evolution.[7] Given the myriad factors documented to affect warfighting evolution, we limit our universe of cases to those proven to facilitate warfighting evolution. We cannot distinguish between good interface designs and better designs if both fail to foster innovation because of the lack of a strong champion or because the threat environment has changed in a way that renders the standard obsolete. Furthermore, finding paired cases in which the organizational and environmental factors are sufficiently similar that they would not confound the interface design effects is exceedingly difficult. For example, if design A led to success and design B led to failure, there would still be a near-infinite number of other contextual variables that could be causal of system B's failure to evolve. By examining success cases only, we hope to illuminate a range of paths that could lead to evolutionary success. Having understood a range of paths, later research might focus on factors that enable specific paths.

We also selected cases from the fairly recent past. Selecting recent cases improves our access to engineers from the original interface design team or from the evolutionary design team to obtain firsthand information about the design factors that might have helped or hindered evolution. Furthermore, software development practices change rapidly. Selecting cases from the relatively recent past of software development practice might help us distinguish factors related to the design of interfaces as opposed to the practices of interface and system development.

Bureaucratic and Programmatic Success in Government, Cambridge, Mass.: Harvard University Press, 1972; Owen R. Coté, *The Politics of Innovative Military Doctrine: The U.S. Navy and Fleet Ballistic Missiles*, dissertation, Boston, Mass.: Massachusetts Institute of Technology, 1996; Barry R. Posen, *The Sources of Military Doctrine: France, Britain, and Germany Between the World Wars*, Cornell Studies in Security Affairs, Ithaca, N.Y.: Cornell University Press, 1986; Jon Schmid and Jonathan Huang, "State Adoption of Transformative Technology: Early Railroad Adoption in China and Japan," *International Studies Quarterly*, Vol. 61, No. 3, September 2017; Jon Schmid, *The Determinants of Military Technology Innovation and Diffusion*, dissertation, Atlanta, Ga., Georgia Institute of Technology, 2018a; Jon Schmid, "The Diffusion of Military Technology," *Defence and Peace Economics*, Vol. 29, No. 6, 2018b; and Jon Schmid, "Intelligence Innovation: Sputnik, the Soviet Threat, and Innovation in the US Intelligence Community," in Margaret E. Kosal, ed., *Technology and the Intelligence Community*, New York: Springer International Publishing, 2018c.

[6] T. M. Egyedi, "Institutional Dilemma in ICT Standardization: Coordinating the Diffusion of Technology," in Kai Jakobs, ed., *Information Technology Standards and Standardization: A Global Perspective*, Hershey, Pa.: IGI Global, 2000; and Ole Hanseth, Eric Monteiro, and Morten Hatling, "Developing Information Infrastructure: The Tension Between Standardization and Flexibility," *Science, Technology, and Human Values*, Vol. 21, No. 4, Fall 1996.

[7] Richard Swedberg, "Exploratory Research," in Colin Elman, John Gerring, and James Mahoney, eds., *The Production of Knowledge: Enhancing Progress in Social Science*, Cambridge, United Kingdom: Cambridge University Press, 2020; and Robert K. Yin, *Case Study Research and Applications: Design and Methods*, 6th ed., Thousand Oaks, Calif.: Sage Publications, October 2017.

Ultimately, we selected two cases: Standardization Agreement (STANAG) 4607 and Military Standard (MIL-STD) 6016.[8] STANAG 4607 establishes the North Atlantic Treaty Organization (NATO) ground moving target indicator (GMTI) data format used for the distribution and analysis of radar imagery for detecting the movement of ground personnel, trucks, convoys, and other assets. MIL-STD-6016 is a messaging standard designed to be transported using a secure, jam-resistant, high-capacity, wireless tactical data link (TDL) colloquially referred to as *Link 16*. These cases were selected based on their documented success in enabling warfighting evolution, the endurance of the standards themselves, and the relative availability of data.[9] Although Link 16 did not originate in the modern era of software development practice, it is actively maintained using modern practices.

Data Sources

In building the case studies, we relied on four primary sources of data: secondary sources describing various aspects of the cases, the technical literature documenting warfighting evolution, interviews with experts, and the standards themselves.[10] Secondary sources included a history of the MITRE Corporation, which had a history of Link 16 terminal development,[11] and a Northrop Grumman publication on voice and data links that included a description of Link 16 technical characteristics.[12] Technical documents were used to identify instances of warfighting evolution involving the standards under scrutiny. To find these documents, we queried publication and patent databases in search of instances in which a standard was used in the service of warfighting evolution.[13]

To attain firsthand accounts of the development and implementation of the standards in practice, we conducted semistructured expert interviews. In constructing our sample of interview subjects, we sought to ensure diversity regarding the subject's relationship vis-à-vis the standard in question. To this end, we interviewed individuals from a wide range of functional roles, including third-party developers, engineers who developed the standards, and former program managers responsible for integrating the standard onto major military platforms. To attain firsthand data regarding the governance and administration of the standard,

---

[8] STANAG 4607, *NATO Ground Moving Target Indicator (GMTI) Format*, Brussels, Belgium: North Atlantic Treaty Organization Standardization Agency, September 2010; and Army Training and Doctrine Command, *TADIL J: Introduction to Tactical Digital Information Link J and Quick Reference Guide*, Fort Monroe, Va.: Defense Technical Information Center, 2000. Ideally, we would have relied on the most recent version of MIL-STD-6016. However, the most recent version of the standard is not publicly available. Therefore, we rely on this guide's interviews to describe the overall structure of the standard.

[9] A careful reader may object that to select cases based on success in facilitating warfighting evolution is to select for the dependent variable, which is a form of selection bias. This bias prevents sound inference-making because the studied sample is not representative of the target population. However, given that the objective of our research is not to make inferential claims, but rather to identify a set of interface factors that warrant further study, this concern does not apply. For a discussion of case study design and inference logic, see Jack S. Levy, "Case Studies: Types, Designs, and Logics of Inference," *Conflict Management and Peace Science*, Vol. 25, No. 1, 2008.

[10] For this report, we interviewed a variety of experts with experience in the domains under study. Interviews were conducted remotely with experts with firsthand experience drafting, amending, or implementing the standards in question. The interviews took place between January 2021 and March 2021.

[11] Davis Dyer and Michael A. Dennis, *Architects of Information Advantage: The MITRE Corporation Since 1958*, McLean, Va.: Community Communications Corp., 1998.

[12] Northrop Grumman, *Understanding Voice and Data Link Networking: Northrop Grumman's Guide to Secure Tactical Data Links*, San Diego, Calif., No. 135-02-005, December 2014.

[13] The data sources queried were the Web of Science website and the Google Scholar search engine for publications and the Derwent Innovation Index research tool and the Google Patents search engine for patents.

including how the standards have evolved to accommodate novel capabilities, we interviewed members of the standards' custodial support teams, various individuals who participated in writing or amending the standard, and standard custodians.

Our final data sources are the standards themselves. The full text of STANAG 4607 and the STANAG 4607 implementation guide are publicly available. MIL-STD-6016 contains restrictive data and thus is not publicly available. Therefore, our primary reference for Link 16 is a TADIL J quick reference guide, an open-access document that includes the basic structure of the Link 16 message standard.[14]

Organization of This Report

In Chapter Two, we discuss the results of our literature review regarding how the design and governance of interfaces might enable capability evolution. Although primarily drawn from the commercial sector, we offer insights on aspects that may be applicable in a DoD environment. Chapters Three and Four cover the two selected cases in depth. For each case, we provide a brief history of the standard, an overview of the standard's technical features, a depiction of the standard's governance process, and a description of novel military capabilities enabled by the standard. In Chapter Five, we present our results: a set of design and governance features that appear to have promoted the identified evolutions in military capability.

---

[14] Army Training and Doctrine Command, 2000.

CHAPTER TWO

Technical Standards and Innovation

Technical standards are documented agreements containing technical guidelines to ensure that materials, products, processes, representations, and services are fit for their intended purpose.[1] In the free market, standards have historically emerged through competition: firms develop their own technology and standards in-house and compete against other firms' standards to capture market share. Examples of this are common and include the QWERTY keyboard, videocassette recorders (VCRs), compact discs, personal computers (PCs), and high-definition television.[2] This approach, which creates a de facto standard, imposes significant costs to losing firms but may yield a desirable outcome for certain vendors and end users. Yet, in the context of de facto standards, the costs of stranded technologies can be large.[3] Even in the free market, firms often prefer to collaboratively draft standards through official standards bodies or committees, a process that leads to a de jure standard.[4]

Each variant carries strengths and weaknesses. In general, *de facto standards* tend to create monopolies, but they initially facilitate innovation because they are responsive to consumer demand. *De jure standards* more evenly distribute economic gains by creating fewer losers but might hamper innovation because of the lack of direct consumer feedback.[5]

Because they are established and maintained by a centralized standard setting body, the standards considered in Chapters Three and Four of this study are de jure standards. However, given the idiosyncrasies of the military technology ecosystem, they are de jure standards that are embedded in a unique market structure. In the context of military interoperability standards, the U.S. government is a monopsonist (i.e., a single buyer with strong market power) with substantial control over the suppliers permitted to enter the market. These conditions allow the U.S. military to simply select or create a winning standard and impose its use

---

[1]   This definition is adopted from Robert H. Allen and Ram D. Sriram, "The Role of Standards in Innovation," *Technological Forecasting and Social Change*, Vol. 64, No. 2–3, 2000.

[2]   Peter Grindley, *Standards, Strategy, and Policy: Cases and Stories*, New York: Oxford University Press, 1995.

[3]   DoD has often been the victim of a stranded technology. As Frey notes about the Defense Information Infrastructure Common Operating Environment (DII COE), "the opinion of most program managers and technical staff is that DII COE standards are already outdated and commercial standards are more useful for improving interoperability" (see Stephen E. Frey, *A Strategic Framework for Program Managers to Improve Command and Control System Interoperability*, thesis, Boston, Mass.: Massachusetts Institute of Technology, 2002).

[4]   Jae-Yun Ho and Eoin O'Sullivan, "Dimensions of Standards for Technological Innovation—Literature Review to Develop a Framework for Anticipating Standardisation Needs," *EURAS Proceedings 2015: The Role of Standards in Transatlantic Trade and Regulation*, 2015.

[5]   Jaesun Wang and Seoyong Kim, "Time to Get In: The Contrasting Stories About Government Interventions in Information Technology Standards (the Case of CDMA and IMT-2000 in Korea)," *Government Information Quarterly*, Vol. 24, No. 1, January 2007.

among vendors via mandate.[6] Yet, a mandated standard is not sufficient to guarantee longevity, acceptance, or evolution.

DoD's history is littered with examples of mandated standards that failed to meet the goals underlying their development. For example, despite begin a well-engineered standard, the DoD mandate to standardize use of the Ada programming language in the 1980s failed to achieve broad uptake outside select DoD programs.[7] A DoD mandate to standardize on the DII COE in the late 1990s was quickly made obsolete when commercial software providers developed superior techniques for providing portability of software applications.[8] Finally, the DoD mandate to standardize on the Software Communications Architecture (SCA) for software-defined radios in the early 2000s failed to garner widespread adoption. Prior RAND research regarding the SCA found the standard to have failed to gain commercial adoption, stating,

> While the government successfully built an influential standard in a new technology area where few competing standards existed and has successfully used the standard to promote waveform level interoperability with allies and partners, there is no way to predict the pace, timing, or scale of commercial adoption. As a result, the hoped-for boost from commercial development has not yet appeared.[9]

Given that mandating a standard's adoption, even in a monopsonistic market, is insufficient to guarantee widespread adoption let alone capability evolution, what can the existing literature tell us about how to design and govern a standard to enable military innovation? The remainder of this chapter focuses on the relationship between standard design, standard governance, and innovation.

Economic Forces: Network Effects and Switching Costs

Two economic factors—network effects and switching costs—have been shown to influence the relationship between standards and innovation. This section defines these factors and describes their observed impact on two technical standards.

Network Effects

A network effect occurs when a product, technology, or standard provides greater utility to both its new and established consumers as it becomes more widely used.[10] Telephones are often used to illustrate network effects. If only a single telephone had been manufactured and sold, it would give no real value to anyone—

---

[6] Note, however, that there is considerable competition between the U.S. armed services that does at times resemble the dynamics among firms. The consumers in this market are the warfighters themselves. Although they do not purchase services and tend to accept whatever products are foisted on them, they do have a vested interest in the effectiveness of the services and products produced.

[7] Some observers of the software industry have noted that DoD's support of Ada may have actually short-circuited the needed evolution in programmer-friendly tooling (e.g., compilers and integrated development environments) that would have made the language more appealing to the larger software development industry (see, for example, Matthew Heaney, "Why Ada Isn't Popular," AdaPower, December 8, 1998).

[8] Frey, 2002.

[9] Bonnie Triezenberg, Graham Andrews, and Padjama Vedula, unpublished RAND Corporation research, 2019.

[10] The term *network effect* is often erroneously called a *network externality* (the latter causes market failure and, as shown in Stan J. Liebowitz and Stephen E. Margolis, "Network Externality: An Uncommon Tragedy," *Journal of Economic Perspectives*, Vol. 8, No. 2, 1994, occurs much less frequently than the former does).

even, and perhaps especially, to its sole user. But, as the telephone consumer-base grows, so too does the number of people one can contact and, simultaneously, the value the product offers to each consumer.

In a combat setting, communications systems that link many warfighters realize network effects as more communications nodes are added to the network. In practice, realized network effects in a combat setting have been smaller than in a commercial setting because the total number of active warfighters is limited, communication security is at a premium, and true peer-to-peer command and control has yet to gain a foothold. Today's U.S. military command and control structures are centrally planned and locally executed, and their small teams operate semiautonomously. Network effects on military tactics, training, and procedures have been the subject of past research.[11]

Switching Costs

Switching costs are the costs incurred by a consumer or technology supplier that are associated with changing from their current product to an alternative. Such costs are often high, particularly in information and communication technology industries where changes must often be percolated through a very large network of devices.[12] Take, for example, consumers who switched their preferred movie player from VHS to DVD in the early 2000s and lost their VHS movie inventories in the process.[13] Similarly, PC users who switch between a Windows versus Mac operating system must learn new keyboard shortcuts, master different software packages, and purchase new peripherals. This effect is also true in warfighting—as teams switch over to a new technology, their readiness to face today's threats can often be adversely affected in the short term even if the longer-term impact of the technology's adoption is to allow them to reach and maintain higher levels of readiness.

The combination of network effects and switching costs cause markets to tip in favor of one technology or standard.[14] In the following sections, we examine two historical cases where these factors had a large impact on innovation.

VCR: Betamax Versus VHS

In the mid 1970s, a new type of video analog recorder—the cassette—was introduced to the home consumer market. These cassettes were designed to replace older home movie technologies with a more durable way to store and a simpler way to replay home movies or, perhaps more crucially, consumer-recorded television

---

[11] In the early 2000s, network-centric warfare became a subject of intense study, and it was hypothesized that it would bring about a "revolution in military affairs." That revolution is quietly occurring, but probably slower than many had envisioned. The Information Age Transformation Series, sponsored by the OUSD-R&E in the early 2000s, is recommended reading for those interested in this topic. In the series, David Alberts and Richard Hayes's 2003 book is particularly applicable to this study (see David S. Alberts and Richard E. Hayes, *Power to the Edge: Command . . . Control . . . in the Information Age*, Washington, D.C.: U.S. Department of Defense, Command and Control Research Program, 2003). All of the books in the series were published by the Command and Control Research Program and are now available online (see International Command and Control Institute, homepage, undated).

[12] Dong-Hee Shin, Hongbum Kim, and Junseok Hwang, "Standardization Revisited: A Critical Literature Review on Standards and Innovation," *Computer Standards and Interfaces*, Vol. 38, February 2015.

[13] This example comes from David Dranove and Neil Gandal, "The Dvd-vs.-Divx Standard War: Empirical Evidence of Network Effects and Preannouncement Effects," *Journal of Economics & Management Strategy*, Vol. 12, No. 3, 2003.

[14] Joseph Farrell and Paul Klemperer, "Chapter 31: Coordination and Lock-In: Competition with Switching Costs and Network Effects," in M. Armstrong and R. Porter, eds., *Handbook of Industrial Organization*, Vol. 3, Amsterdam: North Holland Publishing, 2007, pp. 1967–2072.

shows.[15] Sony, with its Betamax, was the first to bring a videocassette to market. As the first market entrant, Sony hoped to make its format the standard in the industry; however, JVC was not far behind and brought a competing format to market within a year called the video home system (VHS). In the Betamax-versus-VHS standards war, firms were essentially competing on video quality (where Betamax had the edge) versus length of recording (where VHS had the edge). Although JVC was the second mover and had lower-quality video, it was highly competitive on price and recording duration.[16] Perhaps more important to winning the overall standards war, JVC's licensing fees were lower than Sony's. These lower licensing fees allowed manufacturers to produce the format at a lower cost.[17] Overall, this approach allowed JVC to rapidly grow VHS's user- and vendor-base, realizing positive network effects.

In his analysis of this case and others, Grindley concludes that a sound standards strategy requires a firm—or in the case of UC2, requires DoD—to build an installed base faster than its rivals and establish credibility with influential early adopters. Both strategies spur the network effects that perpetuate consumers' demand.[18] Perhaps the most pertinent lesson learned from this example is that a well-engineered standard will be competitive, but a serviceable one that exploits network effects can best it.

The exploitation of network effects directly benefits users and encourages broader uptake of the technology, but it also likely has larger, indirect benefits. A larger user base generates more feedback than a smaller one and helps establish a more active user community, which, in turn, incentivizes use and guides future innovation priorities for standard committees.

Keyboards: DSK Versus QWERTY

High switching costs and network effects make incumbent standards difficult to dislodge even when faced with a superior alternative. An example that illustrates the power of switching costs to create path dependencies is the Dvorak Simplified Keyboard (DSK) and its failed attempt to dethrone the QWERTY keyboard layout.[19] In the late 1860's, Christopher Lathan Sholes invented the QWERTY layout for keys on a typewriter. One of the explanations for why he arrived at the QWERTY layout of keys is that in its earliest form the typewriter was prone to typebar clashes—a common error where a typebar became "stuck" and subsequent keystrokes simply repeated the "stuck" typebar's input. It is hypothesized that the QWERTY layout slowed typists down enough to eliminate the typebar clash. The layout may also have been influenced by telegraph operators who understood that placing commonly used letters further apart would lead to faster, more rhythmic and more error free typing. Whatever its origin, and despite the fact that advances to typewriters' mechanical design solved the typebar clashing problem by the 1880's, the QWERTY layout remains the dominant keyboard layout for English speaking consumers, having made the switch to computer keyboards in the 1960s.

Many alternatives to the QWERTY layout exist. The most common alternative is the DSK layout, which has nevertheless gained little market penetration since it was invented in 1936. The DSK layout was explicitly designed to reduce typing errors and fatigue while increasing a typist's speed and is thought by some scholars

---

[15] Although magnetic tape had begun to replace film in professional television production studios much earlier, 1975 was the first year that videotape for consumers was marketed.

[16] Although Sony's prices were higher than JVC's, it is difficult to determine whether JVC was selling at a loss. Manufacturing costs drop when products are produced at scale, making a short-term loss a long-term profit if the standards war is won.

[17] Grindley, 1995.

[18] Grindley, 1995, p. 39.

[19] Paul A. David, "Clio and the Economics of QWERTY," *American Economic Review*, Vol. 75, No. 2, May 1985.

to be a technically superior design.[20] Today, many computer operating systems offer the DSK layout as an option, but it is rarely used. Low adoption of the DSK standard appears to be explained largely by switching costs (i.e., the cost of retraining on a new layout) and network effects (i.e., the benefit of typists being able to use their skill on most other machines).[21]

Perhaps the most pertinent lesson for DoD from the QWERTY-versus-DSK competition is that incumbent standards with large network effects and high switching costs are hard to unseat, even with superior alternatives. A second lesson is that future operational concepts—such as that experienced when PCs made the occupation of typist obsolete—can dramatically change what matters to warfighter effectivity. Before the introduction of the PCs, the exercise of writing was largely independent of the exercise of typing; content creators dictated or handwrote their thoughts and typists transcribed them. Today, writing and typing are intertwined and other factors besides keyboard layout dominate the output quality of our efforts. Additional keystroke efficiencies are of reduced benefit when the typist is also the content creator.

Governance of Standards Bodies

In addition to network effects and switching costs, the standard governance process has been shown to affect standards and their impact on innovation. This section describes findings on standard governance and then considers how these principles apply in practice by considering the role of governance in the competition between International Organization for Standardization (ISO) and Internet Engineering Task Force (IETF) to standardize the information infrastructures behind today's internet packets and protocols.

In his 2017 technical paper, Schmidt details the anatomy of standards bodies and issues recommendations to organizations attempting to push their standards through a formal committee.[22] Although our research does not necessarily lead us to recommend a formal standards committee for the UC2 effort, Schmidt's description of committees and their members holds valuable insights that DoD should consider when developing governance practices. Key considerations are detailed as follows.

Members of standards committees generally include individuals, companies, government agencies, or academic institutions. The process to obtain membership is either open or restricted by fee, nomination, or appointment. As will be observed in our discussion of the internet standards, eligibility for membership can have a large impact on the standard that is produced.

The process to propose and finalize design decisions and make amendments is also an important determinant of a standard's eventual content. Most committees achieve consensus through formal votes or via working groups working toward a final design decision. In general, our research leads us to believe that iterative processes that encourage feedback may be more effective at producing flexible standards. Broad and diverse representation demands more than broad and diverse membership; it requires a process by which all members can voice concerns or propose alternative solutions.

Standardization through committee is as much a diplomatic process as it is a feat of collective engineering. Winning standards do not emerge on their technical merits alone. A committee participant's reputation, soft skills, and ability to persuade affect that person's ability to garner support for a technical conces-

---

[20] It is worth noting that the inferiority of the QWERTY design is contested by Liebowitz and Margolis, who claim that the evidence for the vast technical superiority of the DSK layout is weak and overstated (see Stan J. Liebowitz and Stephen E. Margolis, "The Fable of the Keys," *Journal of Law and Economics*, Vol. 33, No. 1, April 1990). The standard view is most convincingly described in David, 1985.

[21] Farrell and Klemperer, 2007.

[22] Charles M. Schmidt, *Best Practices for Technical Standard Creation*, McLean, Va.: MITRE Corporation, technical paper, 2017.

sion. Although the politicization of de jure standardization is inevitable, committee leaders may be wise to minimize its influence; the objective is, after all, to produce a technically competent standard that performs its necessary functions and suits the demands of suppliers and users alike. If parochial interests dominate, the standard's quality might suffer. The risk is non-negligible: Practitioners and observers have increasingly voiced disappointment in the "technical content of new standards and the processes of committee standardization."[23] Too often, standardization efforts become a "complex contest between corporate wills"—an outcome that occurs more frequently when committee members have preestablished technologies.[24] Unfortunately, solutions that disincentivize politicking are scarce. Schmidt suggests politicization may be, in part, reduced by forcing much of the discussion into official channels and discouraging backroom deals. As will be shown below, IETF's insistence of working code and rough consensus seems to us a better approach. In either case, a committee benefits from leaders who are knowledgeable—allowing them to distinguish contributions that improve the standard from those that serve only the interests of a specific participant or faction—and persuasive, leading by influence rather than authority.

Two other factors that affect the success of a standards committee are an understanding of the standard's intended market and "the willingness of firms to commit written technological contributions to the standard committee."[25] That an understanding of consumer demand is crucial has been detailed in the use cases throughout this chapter, as has its solution: The standards committee must broadly understand and represent the stakeholder's interests. Firms will be willing to commit written technological contributions only when they believe the value of standardization outweighs the risks and that contributions have a fair chance of influencing the standard.

The Internet: ISO Versus IETF Standards

The effect of standard design and governance on standard outcomes is particularly evident in the standards competition between the ISO and IETF to determine the information infrastructures behind today's internet packets and protocols. In 1996, Hanseth and colleagues studied the competing effects of standardization on innovation—or, in their words, flexibility—within the context of this standard competition.[26] In particular, they considered two variables: the layering abstraction on which our modern networks are built and the governance structures of the ISO versus those of the IETF. In prior RAND research for OUSD-R&E FNC3, we described the difference in network-layering abstractions between the ISO standard and the informal internet abstractions used by IETF members (see Figure 2.1). We also reproduce the description here verbatim because it provides critical context to the governance implications of the study from Hanseth, Monteiro, and Hatling, 1996:

> Over the past 20 years, great strides have been made in achieving technical interoperability of computing systems. A key enabler has been the abstraction of the communication between any two nodes in the system as a *stack* or set of layers that provide the functions required for any two systems to exchange information. The two most commonly referenced abstractions are the Open Systems Interconnection (OSI) and internet models of communication.
>
> The OSI model has seven layers and was designed such that each layer would strictly depend only on the services of the layer immediately below it. Although originally developed with a set of accompanying pro-

---

[23] Egyedi, 2000.

[24] Carl F. Cargill, "Why Standardization Efforts Fail," *Standards*, Vol. 14, No. 1, Summer 2011.

[25] Cargill, 2011.

[26] Hanseth, Monteiro, and Hatling, 1996.

FIGURE 2.1

The OSI and Internet Models of Communications

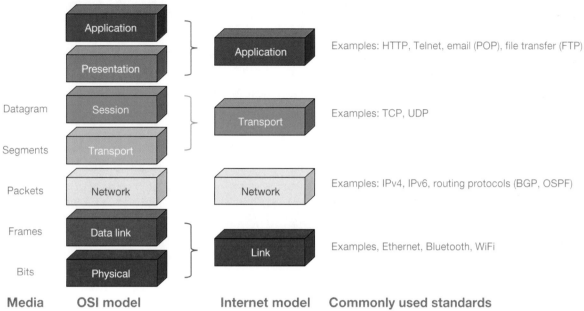

SOURCE: Dimarogonas et al., 2021, Not available to the general public.
NOTE: BGP = border gateway protocol; FTP = file transfer protocol; HTTP = hypertext transfer protocol; OSPF = Open Shortest Path First; POP = post office protocol; TCP = transmission control protocol; UDP = user datagram protocol; WiFi = wireless fidelity.

tocols at each level that would communicate at that layer, the protocols were unwieldy and never caught on. However, the *idea* of the seven OSI layer model persists to this day and provides a common vocabulary for network analysts.[27]

The abstracted internet model is simpler, having only four layers, and is more representative of actual practice. The seven-layer model is contrasted with the seven-layer model in Figure [2.1] and commonly used communication standards (i.e., *protocols*) and the media of that communication are roughly mapped against them.[28]

Despite their technological similarities, the OSI and the IETF differed greatly in their approach to standardization. OSI was the product of the ISO, a formal international standards body, where nations receive a predefined number of votes and propose ideas for protocols using structured processes. The committee designing the OSI reportedly paid little attention to implementation or compatibility with non-OSI prod-

---

[27] The model can be purchased from ISO (see ISO and the International Electrotechnical Commission, "Information Technology—Open Systems Interconnection: Basic Reference Model: The Basic Model—Part 1," ISO/IEC 7498-1:1994, webpage, 1994). However, hundreds and perhaps thousands of representations of the model can be found online using the term *OSI 7 layer model* in any search engine.

[28] James Dimarogonas, Jasmin Leveille, Jan Osburg, Shane Tierney, Bonnie Triezenberg, Graham Andrews, Bryce Downing, Muharrem Mane, and Monica Rico, *Universal Command and Control Language Early Engineering Study, Performance Effects of a Universal Command and Control Standard*, Santa Monica, Calif.: RAND Corporation, 2021, Not available to the general public.

ucts; instead, local technical considerations dominated.[29] The IETF's committee and standardization process, however, are much different.

The IETF is deliberately informal and has open membership, which engenders broad and diverse representation. The IETF process emphasizes prototyping, evolutionary deployment, learning, and user involvement.[30] Within IETF, standards are developed through a three-phase process known as the request for comments (RFC) process: proposed standard, draft standard, and final standard. To become a draft standard, at least two independent implementations must be developed and evaluated. To become a final standard, the implementations must become a clear de facto standard, meaning that many of the most-influential RFCs spend years as draft standards. By pitting independent proposals, drafts, and implementations against one another, the IETF's standardization approach draws from the best parts of the de facto process (i.e., competition) while minimizing the stranding costs that can characterize standardization via market competition.

Technical design paradigms used in the IETF and OSI technical standards also add to the flexibility of the standards. Both rely upon layering and black-boxing to introduce flexibility into their technologies, although neither of these principles are strictly adhered to in either set of standards. The overall goal of this design approach is to allow features to be added without interfering with existing ones. The internet has both diffused widely and changed rapidly. Hanseth, Monteiro, and Hatling state that "this [diffusion and rapid change] is possible, among other reasons, because it [is] based on a format defined in such a way that any implementation may simply skip or read as plain text those elements that it does not understand. In this way, new features can be added so that old and new implementations can run together."[31]

Another design principle essential to successful evolution of internet standards is the separation of concerns. A cautionary tale can be found in the OSI's application-level standard for email. In the standard, the OSI combined the tasks of uniquely identifying a person with the task of implementing the way a person is located. The downside of this approach is that if an organization were to change the way its email system locates a person (e.g., by changing its network provider), the unique identifiers of personnel in the organization would have to be changed as well. The ISO's email standard is rarely used in today's internet. This example demonstrates how failure to properly segregate concerns increases switching costs and, with it, the costs to subsequent innovation.

Although the IETF standards have proven to be more agile than the OSI standards, they have fallen victim to one of the unique challenges of networked systems: scalability. The original internet protocol packet format used a 32-bit numeric address to uniquely identify the destination and source of packets. By the mid-1990s, it became apparent that the internet's exponential growth would soon exhaust all possible addresses. To resolve this problem, the IETF changed the internet protocol packet format to include a 128-bit alpha-numeric address and a new protocol—the Neighbor Discovery Protocol (NDP)—to map the media access control addresses that are unique to all devices connected to the internet.[32] This RFC entered as a proposed standard in 1995 and achieved draft standard status in 1998. In keeping with the IETF's philosophy

---

[29] Marshall T. Rose, "The Future of OSI: A Modest Prediction," *ULPAA '92: Proceedings of the IFIP TC6/WG6.5 International Conference: Upper Layer Protocols, Architectures and Applications*, Vol. C-7, 1992.

[30] Douglas Schuler and Aki Namioka, eds., *Participatory Design: Principles and Practices*, Boca Raton, Fla.: CRC Press, 1993; and Paul Hoffman and Susan Harris, *The Tao of the IETF: A Novice's Guide to the Internet Engineering Task Force*, RFC 4677, September 2006.

[31] Hanseth, Monteiro, and Hatling, 1996.

[32] The revised standard, termed *IPv6* to distinguish it from the prior *IPv4* standard, has several other small changes with respect to header and checksum fields, but the primary changes are the expanded address field and the NDP. A summary of IPv4 versus IPv6 can be found at the Guru99 website (see Lawrence Williams, "IPv4 vs IPv6: What's the Difference Between IPv4 and IPv6?" Guru99, webpage, last updated October 7, 2021). Currently, many computers have the capability to run both IPv4 and IPv6 in what is called a *dual stack configuration*.

of not declaring a standard to be final until it has clearly become a de facto standard, it was not ratified as a final standard until 2017. Although the committee foresaw and solved the impending address constraint, this example highlights the importance of scalability in the design of technical standards.

Perhaps the most important lesson for DoD from the ISO-versus-IETF competition is the large influence of standard governance on outcomes. A less formal organization of volunteers with a clear preference for "rough consensus and running code" over formal processes and voting has successfully managed the network technology that has changed the world.[33] Their processes are deliberately designed to balance the best of de jure and de facto standards evolution in a way that minimizes the expense and probability of stranding technologies while allowing for transparency into the change process so that users can properly manage, predict and prepare for the inevitable switching costs that arise when a new standard prevails. It seems unlikely that the IETF's processes would be fully adopted in the hierarchical and formal authority culture that characterizes DoD. Yet, a process that has successfully navigated the competing interests of such large corporations as Cisco, Huawei, IBM, and Juniper Networks to produce generally well-regarded technical standards might offer lessons for managing the competing interests of the U.S. Navy, U.S. Army, U.S. Air Force, U.S. Space Force and the myriad agencies within the DoD acquisition and operational communities.

How Does Standardization Affect Innovation? Summary of Lessons from the Standardization Literature

The research and historical examples reviewed in this chapter supply a few clear principles regarding conditions that may allow DoD to enjoy the benefits of network effects, minimize the effects of switching costs, and avoid technology lock-in.[34] They are as follows:

- Build, early in the process, an engaged warfighter and contractor base to ensure the standard is solving the right problem and balances the needs of both.
- Leverage network effects by adding users. For most DoD interoperability standards, the value of a standard increases as nodes are added to the network.
- Design standards that are both flexible and extensible enough to adapt to future operational concepts.
- Favor rough consensus and working code over formal processes. When in place, formal processes should be open and transparent.

---

[33] Hoffman and Harris, 2006, p. 5, summarizes the governance philosophy thusly: "In many ways, the IETF runs on the beliefs of its members. One of the 'founding beliefs' is embodied in an early quote about the IETF from David Clark: 'We reject kings, presidents and voting. We believe in rough consensus and running code.'"

[34] Although lessons derived from commercial standards might unearth variables that are likely to be important to successful standard design and governance in a military setting, implementation of commercial best practices into a government setting requires careful accounting for the particular conditions into which these practices are being implemented (Jeffrey A. Drezner, Jon Schmid, Justin Grana, Megan McKernan, and Mark Ashby, *Benchmarking Data Use and Analytics in Large, Complex Private-Sector Organizations: Implications for Department of Defense Acquisition*, Santa Monica, Calif.: RAND Corporation, RR-A225-1, 2020).

CHAPTER THREE

STANAG 4607: NATO's Standard for GMTI Data

Overview of STANAG 4607

STANAG 4607 establishes the NATO GMTI data format. Its principal objective is to "promote interoperability for the exchange of ground moving target indicator radar data among North Atlantic Treaty Organization (NATO) intelligence, surveillance, and reconnaissance (ISR) systems."[1] To this end, the standard defines the format for transmitting raw GMTI data generated by a sensor (e.g., an airborne surveillance radar) to a downstream exploitation system capable of transforming the data into usable information (e.g., for use in targeting).

History of STANAG 4607

One of the primary impetuses for the development of STANAG 4607 was a desire to fulfill user demand derived from military operations.[2] Although still under development, the Joint Surveillance Target Attack Radar System (STARS), an airborne platform for conducting ground surveillance, battle management, and command and control, was deployed during the 1990–1991 Gulf War. At this point, Joint STARS used a proprietary data format standard for the transmission of GMTI data that had been developed by the Joint STARS's prime contractor.[3] This proprietary standard satisfied a narrow GMTI requirement; it allowed the low-bandwidth transmission of GMTI data to the Army Common Ground Station (CGS) system. Following the initial deployment of Joint STARS into operation, there was dissatisfaction with the proprietary GMTI standard due to its narrow technical scope (i.e., low-bandwidth transmissions to a fixed number of Service-specific platforms) and the effect of the propriety standard in limiting data access. The desire to broaden the technical scope beyond low-bandwidth transmissions and to more fully and widely exploit GMTI data created demand for a new, forward-looking, open GMTI standard.[4]

Besides creating overall demand for a new and open GMTI standard, demand from operators also played a role in determining the eventual technical character of STANAG 4607. In particular, operators during the 1990–1991 Gulf War expressed a desire to use GMTI data to identify and target moving ground vehicles, such as the transporter erector launcher (TEL) vehicles that were used to transport and launch Iraqi Scud

---

[1] The first edition of the standard was promulgated on March 11, 2005.

[2] This section relies heavily on the account provided in Clem H. Huckins, *Evolution of a Standard: The STANAG 4607 NATO GMTI Format*, McLean, Va.: MITRE Corporation, technical paper, February 2005. Huckins was the technical lead and chief editor of STANAG 4607 at the time that the 2005 document was written.

[3] The prime contractor was Northrop Grumman. Authors' interview, January 4, 2021.

[4] Authors' interview, January 4, 2021.

missiles. However, this kind of time-sensitive targeting requires a more sophisticated use of GMTI data than was available in 1991. This demand contributed to the inclusion of a high-range resolution (HRR) message capacity within what would eventually become STANAG 4607.[5]

In 1998, the Assistant Secretary of Defense for Command, Control, Communications and Intelligence (ASD/C3I) charged the U.S. Air Force to develop a common GMTI data standard to replace the various, often proprietary, GMTI message standards used at that time.[6] To this end, the U.S. Air Force stood up an integrated project team (IPT) in May 1999.[7] The IPT sought to develop a common GMTI data standard—initially called the common ground moving target indicator (CGMTI) format—for use across all U.S. military service branches, intelligence agencies, and allied nations.[8]

Initially independent from the U.S.-led initiative, NATO Air Group 4—then the NATO Air Group responsible for ISR—established the NATO GMTI technical support team (TST) in April 2000. The GMTI TST consisted of government and vendor representatives and was charged with evaluating possible GMTI data formats and proposing a common GMTI data standard via a novel STANAG or an addendum to an existing STANAG.

Upon learning of U.S.-led CGMTI effort, NATO TST members began participating in CGMTI IPT meetings.[9] Eventually the efforts were merged into the NATO GMTI TST, which was charged with establishing a NATO STANAG for GMTI data transmission. The TST, led by Clem Huckins of the MITRE Corporation, consisted of government and vendor representatives from the United States, United Kingdom, Canada, France, Germany, and Denmark and representatives from the NATO Consultation, Command and Control Agency.

The decision to transition the U.S.-led CGMTI effort to a NATO standard was also driven by the anticipated proliferation of U.S. and allied platforms. During an interview for this study, a senior engineer with a significant role in shaping the standard described how the platform investment pipeline underscored the utility of developing a NATO standard, stating, "There was an investment roadmap with all of these emerging platforms, for instance, the [Lockheed P-3 Orion], Reapers, Predators—and trying to get everyone on the same page when it comes to GMTI data was the original goal [of moving to a NATO standard]."[10]

During the initial drafting of the GMTI data standard, the TST established three subgroups: Coordinate Systems and Time Standards, HRR, and Structure and Definitions. The Structure and Definitions subgroup was the lead subgroup and was charged with the overall drafting, management, and editing of the final document. The TST, and its comprising working groups, began the standard development process by cataloging the data elements that would be required by a GMTI data standard. To this end, the TST were briefed by various program officers on the technical parameters of the systems that would eventually transmit and receive GMTI data. The TST then surveyed and assessed available legacy standards for their applicability to the cataloged data elements. Finally, the TST drafted the standard to include each of the identified required data elements, leveraging legacy standards when appropriate.[11]

---

[5]   Authors' interview, January 4, 2021.

[6]   Office of the Secretary of Defense, *Unmanned Aerial Vehicles Roadmap: 2002–2027*, Washington, D.C.: U.S. Department of Defense, 2002.

[7]   Huckins, 2005.

[8]   Huckins, 2005.

[9]   Huckins, 2005.

[10]   Authors' interview, January 20, 2021.

[11]   Huckins, 2005.

In November 2002, STANAG 4607 was presented to the NATO ISR Interoperability Working Group to begin the ratification process. In October 2004, STANAG 4607 was tested during a technical interoperability experiment conducted in The Hague under the auspices of the Coalition Aerial Surveillance and Reconnaissance (CAESAR) project.[12] During the test, several minor problems were observed with the standard. For example, it was found that the standard contained conflicting descriptions of coordinate representations. However, these errors were corrected in updated editions of the STANAG 4607, and it was concluded that "STANAG 4607 was found useful, applicable, but not perfect and with room for improvements."[13] Edition 1 of STANAG 4607 was promulgated on March 11, 2005.

Following the initial promulgation of the standard, the TST was designated as the custodial support team (CST) for the standard. The CST later amended the standard to accommodate novel sensor types, including the spaceborne platforms. Edition 2 of STANAG 4607 was promulgated on August 2, 2007. Finally, the current, in-force version of the standard—edition 3 of STANAG 4607—was promulgated on September 14, 2010.[14] Table 3.1 depicts major events in the history of STANAG 4607.

TABLE 3.1

Major Events in the STANAG 4607 GMTI Format

Date	Event
1990–1991	Joint STARS deployment during the 1990–1991 Gulf War underscores demand for an open GMTI standard to allow the use of GMTI data for time sensitive targeting
Mid-1990s	GMTI community of interest is established
1998	ASD/C3I charges U.S. Air Force to develop a common GMTI data format
May 1999	U.S. Air Force establishes IPT to begin work on GMTI standard
April 2000	NATO Air Group 4 establishes TST to begin work on GMTI standard
~2000	IPT and TST merge to form a NATO TST to develop a GMTI data format STANAG
November 2002	STANAG 4607 approved by NATO ISR Interoperability Working Group
2003	NATO Standardization Agency submits draft standard (dated March 2003) to nations for ratification
October 2004	STANAG 4607 successfully tested during a technical interoperability experiment
March 11, 2005	STANAG 4607 (edition 1) promulgated
2005–2007	STANAG 4607 CST amends standard to include new sensor types, including spaceborne platforms
August 2, 2007	STANAG 4607 (edition 2) promulgated
September 14, 2010	STANAG 4607 (edition 3) promulgated

---

[12] Thomas Kreitmair, Joe Ross, and Trond Skaar, *Experimentation Activities with Aerospace Ground Surveillance*, The Hague, Netherlands: NATO Consultation Command and Control Agency, AD-A460950; X5-NATO/C3/NL, June 2005.

[13] Kreitmair, Ross, and Skaaar, 2005.

[14] STANAG 4607, 2010.

Governance of STANAG 4607

The formal governance structure and related management processes for STANAG 4607 is set forth in the standard's implementation guide. Six entities play a role in the governance of STANAG 4607: the Custodian, custodial support team (CST), national representatives, an administrative support team (AST), the NATO Secretary, and ad hoc special teams.[15]

The STANAG 4607 Custodian is the central and most powerful entity in the governance of the standard. The Custodian is responsible for serving as chairman for all CST and configuration management meetings, tracking changes to the standard, maintaining an official version of the standard, tasking the AST, and reporting to NATO parent organizations on the status of the standard.[16] Clem Huckins was the first Custodian of the standard. Following Huckins were Capt Andrew Calhoun of the U.S. Air Force, Bryan Blank of the National Geospatial Agency (NGA) and Brian R. O'Hern of the Air Force Research Laboratory (AFRL). O'Hern became the current STANAG 4607 Custodian in May 2019.

The STANAG 4607 CST is responsible for deliberating and approving any changes or extensions to the standard. In practice, the CST has also been responsible for drafting amendments to the standard.[17] The process for adding an extension to STANAG 4607 is highly centralized and structured.[18] To make an official change to the standard, the originator of the extension must first submit a change request form to the CST. The form requires the originator to provide a description and rationale of the proposed change. The proposal is reviewed by the CST and the Custodian and either approved for further action or disapproved. If the proposal is approved for further action, the proposed extension is added to the compendium of registered extensions for the standard and the proposed extension becomes a registered extension. The Custodian then assigns the extension a name and an identification segment with its own header segment. Following this step, users are permitted to use the extension during implementation, test, and validation (and results of this testing can be submitted to the CST and Custodian for validation). Following successful validation, the extension receives a controlled designation and is added to the Registry of Controlled Extensions for STANAG 4607; at this point, the extension becomes part of the official standard.[19] Finally, any STANAG or STANAG amendment must be translated into a second language before promulgation.

The CST is composed of STANAG 4607 national representatives. Each NATO member nation may designate one representative to the CST. National representatives are responsible for defining the means of determining a nation's position on proposed changes, articulating this position, and serving as the official spokesperson for any other participants from that nation.

Although the exact cadence of CST meetings depends on various factors, including the demand for changes to the standard, the CST meeting schedule and the amendment process have been, in practice, deliberate. During the preparation (approximately from 2005 to 2007) of edition 2 of the standard, the CST met every three to six months. The location of these meetings would rotate between NATO member countries.[20] CST meetings for STANAG 4607 were typically co-located with CST meetings for STANAG 7023 (Air Reconnaissance Primary Imagery Data Standard), STANAG 4609 (Digital Motion Imagery Standard), STANAG

---

[15] Allied Engineering Documentation Publication Number 7 (AEDP-7), *NATO Ground Moving Target Indicator (GMTI) Format, STANAG 4607 Implementation Guide*, Brussels, Belgium: NATO Standardization Agency, May 2013.

[16] AEDP-7, 2013, p. K-2.

[17] Authors' interview, January 8, 2021.

[18] AEDP-7, 2013, p. M-2.

[19] The Registry of Controlled Extensions for STANAG 4607 is contained in the implementation guide for the standard (AEDP-7, 2013). No controlled extensions were listed in the most recent implementation guide.

[20] Authors' interview, January 8, 2021.

4676 (Intelligence, Surveillance and Reconnaissance Tracking Standard), and STANAG 4545 (Secondary Imagery Format).[21] Co-locating these CST meetings allowed participants to attend meetings for each of the complementary data formatting standards. Typically, CST meetings would take place over the course of one week, with two to three days dedicated to STANAGs 4607, 4609, and 4676, and one to two days dedicated to STANAG 7023 and STANAG 4545.

It is interesting to note that the deliberateness of the STANAG CST meeting schedule might have had positive consequences. During an interview for this study, an individual describes the effect of a deliberate meeting cadence on the quality of the changes proposed, stating,

> The STANAG process is slow and methodical [. . .] it gives everybody a chance to really think things through, to go back and consult with subject-matter experts in your home country and to really be thoughtful and forward-looking. I think that was key to ensuring the longevity of the standard. This thoughtful process ensured that there weren't any shortsighted decisions that were made. I never heard cases where a decision from a year ago turned out to be the wrong decision.[22]

The final entities within STANAG 4607 governance structure are the AST, NATO Secretary, and ad hoc special teams. The STANAG 4607 AST performs administrative functions associated with the standard, including arranging meeting times, taking minutes of all meetings, and tracking and disseminating recommended changes to the standard. The NATO Secretary is responsible for maintaining configuration management of the standard's webpage. Finally, the Custodian may convene special teams to consider major technical issues associated with the standard, including large upgrades or amendments to the standard.

The STANAG 4607 User Community

Two important features of the STANAG 4607 user community are worth highlighting. The first feature is the STANAG 4607 community of interest. This forum provided an early platform for a diverse set of stakeholders to discuss the character of operator demand and technical feasibility with regard to the emerging GMTI format. The second feature worth highlighting is a GMTI data repository, which broadened data access, which, in turn, enabled capability development.

STANAG 4607 Community of Interest

During the mid-1990s, a GMTI community of interest was established. The impetus for establishing this community of interest was a growing demand for the use of GMTI data and demand to extend the application of GMTI data to novel uses, including time-sensitive targeting and counter-improvised explosive device (IED) operations. The GMTI community of interest met roughly once a month from the mid-1990s through late 2000s and consisted of individuals from the relevant program offices, the U.S. Air Force, the U.S. Army, and NATO member countries. The community of interest received no funding yet maintained high participation.[23]

---

[21] Authors' interview, January 8, 2021.

[22] Authors' interview, January 8, 2021.

[23] Authors' interview, January 4, 2021.

STANAG 4607 GMTI Data Repository

The user community is also characterized by a STANAG 4607–formatted GMTI data repository called Atlantis. This database was populated with data from Joint STARS, Global Hawk, and other platforms, and was accessible on the Secret Internet Protocol Router Network (SIPRNet) to anyone with a need to know.[24] The AFRL built the Atlantis GMTI database in 2002 to provide the laboratory with the data necessary to develop and build tracking and fusion algorithms. The database originally hosted GMTI data in the NATO Exploitation Format data format but then converted its data into the STANAG 4607 format.

Today, the repository is used by a wide variety of organizations to develop novel GMTI-enabled capabilities and is still hosted by AFRL. However, formal GMTI operations were eventually migrated, assisted by funding from DoD's ISR Task Force, to the NGA.

STANAG 4607 Technical Characteristics

The primary purpose of STANAG 4607 is to facilitate GMTI data exchange between radars and their associated exploitation systems. It does not concern itself with either how the data are collected or the means by which the data are distributed. However, it does include a means for requesting surveillance service from the sensor and receiving acknowledgment that the requested surveillance will, or will not, be performed by the sensor.[25]

STANAG 4607 specifies a binary, message-oriented format, organized by packets. Each packet is split into message segments where each segment contains a particular type of information. Each packet contains data from a single radar job.[26] If the data from one radar job exceed the packet size limit, data can be split into multiple packets. Segments can be transmitted in any order, so long as a header segment proceeds the transmitting segment. The segments include job request, job acknowledge, mission, HRR, job definition, free text, test and status, processing history, platform location, and dwell.[27] The information provided in a segment can be tailored based on the end user's information need. For example, the dwell segment might include target reports, and the HRR segment might include scatterer reports.[28]

The standard defines only the application layer of the networking reference model and is independent of transport, network, and link layers. Therefore, it can be used by itself or it can be embedded into other frame-oriented formats (e.g., other ISR STANAGs, such as STANAG 7023 and STANAG 4545). It can also be translated to eXtensible Markup Language (XML) for dissemination.[29] The STANAG 4607 format allows transmission of usable GMTI data to exploitation systems of various levels of sophistication. Simple GMTI systems can use a small portion of the data format via low-bandwidth channels and sophisticated systems can encode the full output data, including HRR information, over wideband channels.[30]

---

[24] Authors' interview, January 4, 2021.

[25] STANAG 4607, 2010.

[26] STANAG 4607, 2010.

[27] AEDP-7, 2013, pp. A-1–A-2.

[28] AEDP-7, 2013, pp. A-1–A-2.

[29] The Air Force's Advanced Battle Management System cloud-based Unified Data Library (UDL) uses JavaScript Object Notation schemas for GMTI and tracks are directly mapped from the STANAG 4607 and 4676 data models. The UDL is intended as a data repository that merges all available sensor data into a single unified environment to support all-domain operations.

[30] Mário Monteiro Marques, *STANAG 4586—Standard Interfaces of UAV Control System (UCS) for NATO UAV Interoperability*, Portugal: NATO Standardization Agency, Science and Technology Organizations, STO-EN-SCI-271, May 1, 2015.

Adoption of STANAG 4607

Since edition 1 was promulgated on March 11, 2005, STANAG 4607 has been widely adopted. It supports a large number of platforms and sensors across the U.S. military service branches and those operated by NATO member countries. In March 2019, the Custodian of STANAG 4607 approved an interoperability testing program for the standard.[31]

At least one NATO, two UK, and ten U.S. platforms have successfully implemented the standard.[32] Although the standard enables identification of more than 50 platforms, not all of these platforms have implemented the standard.[33] In terms of sensor types, 27 distinct sensors are explicitly mentioned in the most recent implementation guide.[34] Specific sensors include the highly integrated surveillance and reconnaissance radar, the airborne stand-off radar systems, Global Hawk radar, AN/ZPY-1 tactical radar, and Defense Advanced Research Projects Agency's Vehicle and Dismount Exploitation Radar system.

STANAG 4607 and Warfighting Evolution

STANAG 4607 stood out from other candidate standards that we considered for study because there is evidence that it enabled the advent of novel GMTI-based capabilities. A radar can output information in many different ways—from raw signal data to fully processed imagery or location measurements. However, lookdown radars that seek to track ground-based targets must deal with a significant amount of background clutter to properly distinguish an object of interest from the environment around it, and much of the information needed to refine finished products is lost when producing fully processed imagery or measurements.[35] Standardizing a format for raw radar data enabled, at least in part, the evolution of warfighting as new methods were developed to exploit that raw data. As the uses of the raw data expanded, so did the standard. Five specific standard evolutions we identified were

1. support for transmission of HRR image chips
2. support for raw data from space-based sensors
3. support for raw data from synthetic aperture radars (SARs)
4. use of GMTI data to counter the adversary's use of IEDs during operations in Afghanistan and Iraq
5. the use GMTI data for intelligence purposes.

Table 3.2 briefly describes novel capabilities enabled, in part, by STANAG 4607. In Chapter Five, we attempt to trace the specific design and governance features in the standards to the advent of these capabilities. In Appendix A, we describe the capabilities in greater detail.

---

[31] Brian R. O'Hern, Custodian, NATO STANAG 4607 (GMTI), Air Force Research Laboratory (AFRL/RIEA), "STANAG 4607 Interoperability Testing," memorandum to David Schneeberger, Chairman, Joint Capability Group for ISR (JCGISR), Rome, N.Y., March 20, 2019.

[32] Email correspondence, February 19, 2020.

[33] Information on supported platforms and sensors comes from the STANAG 4607 implementation guide, which includes a "Platform Type" and "Sensor Type" field. For example, specifying a platform type = 9 indicates that the data are generated from a Predator platform.

[34] Again, the option for designating "other" is available for sensors as it is for platforms.

[35] Processing designed to amplify a signal and reduce its noise removes information that may be critical to downstream analyses. There is critical information in the noise of the raw measurements that must be preserved if we are to exploit GMTI data to the maximum extent possible.

TABLE 3.2

Capabilities Enabled by STANAG 4607

Capability	Role of STANAG 4607 in Evolution or Advent
Supportability of HRR chips	STANAG 4607 includes an HRR message segment to allow transmission of HRR chips, which can be used for target identification.
Use of space-based sensors for GMTI	The STANAG 4607 amendment process was followed to allow, in edition 2 of the standard, transmission of data from spaceborne assets.
Extending application to multichannel SAR	The STANAG 4607 amendment process was followed to allow, in edition 2 of the standard, transmission of SAR data.
Use of GMTI data to track IED attacks	A set of analysis and development cells use standardized GMTI data from the Atlantis data repository to develop counter-IED applications.
Use of GMTI for intelligence purposes	A set of analysis and development cells use standardized GMTI data from the Atlantis data repository to develop group tracking, forensic backtracking, meeting detection, route avoidance, anomaly detection, road detection, and traffic detection algorithms.

CHAPTER FOUR

The Link 16 Communications System

Overview of Link 16

Link 16 is a relatively old communications system designed for use as a secure, jam-resistant, high-capacity, wireless TDL to exchange near-real-time information. Link 16 can be used to transmit air track, text, imagery, and digital voice data.[1] The Link 16 standard encompassed both a messaging standard and a waveform standard that together encompass the application, transport, network, and link layers of the network model of communication. As we will discuss in this chapter, innovation and evolution in warfighting have been enabled, in part, by reworking the standards into something closer to the network model. The Link 16 waveform today can carry internet protocol packets and, when so used, provides a jam-resistant link layer for the networks, transports, and data it carries. Operational uses for Link 16 include surveillance, electronic warfare, mission management, weapons coordination, air control, fighter-to-fighter net, secure voice channels, navigation, positive friendly identification, and network management.[2]

Link 16 is used by dozens of U.S. and NATO military assets, including airborne, ground, and naval platforms that are called Joint Tactical Information Distribution System (JTIDS) units (or Jus).[3] JTIDS is a system that includes several elements that work together to enable the full operational capability.[4] First, there is the data link itself: the underlying waveform. This waveform runs on several different radios, often referred to as JTIDS or Link 16 terminals, which together form the JTIDS network. Each platform has command and control software that handles the Link 16 messages (in MIL-STD-6016 message format), and the information from these messages are displayed on the aircraft tactical display. Breaking out these different elements of the system separately has allowed each to evolve independently and on different timelines. Today, technology insertion can be pursued separately for the message set, the waveform, the radio, and command and control processing system without having to upgrade everything at the same time.

---

[1] A note on terminology is warranted here. *Link 16* is the term used by NATO to describe the tactical data communication system that includes both the message and waveform standards. The United States uses the term *Tactical Digital Information Link J* (TADIL J) to describe the same system. The term *Tactical Data Link* (TDL) has replaced TADIL in official DoD parlance. In this report, we are primarily concerned with the message standard, which we will refer to as *MIL-STD-6016*. When we refer to both the message standard and the waveform standard, we will use the term *Link 16*.

[2] Army Training and Doctrine Command, 2000.

[3] Often, the terms *Link 16* and *JTIDS* are used interchangeably for the tactical data link. Eric V. Larson, Gustav Lindstrom, Myron Hura, Ken Gardiner, Jim Keffer, and William Little, *Interoperability of U.S. and NATO Allied Air Forces: Supporting Data and Case Studies*, Santa Monica, Calif.: RAND Corporation, MR-1603-AF, 2004.

[4] One can think of the different elements of a Link 16 system by using the analogy of human language: The messages are equivalent to words in a language. The terminal is equivalent to a person's vocal cords and mouth, both of which allow humans to modulate sounds, while the waveform is the modulation of sounds in the mouth and vocal cords to form spoken words. Command and control is the brain that makes sense of the words, makes decisions, and takes action.

Scope and Terminology

Throughout this report, the term *Link 16 terminal* will be used to refer to the radios that transmit and receive data to and from a Link 16 network, and the message standard will be referred to as MIL-STD-6016.[5]

History of Link 16

The intellectual foundations of Link 16 are based on an idea of Gordon Welchman to replace point-to-point switchboard-moderated tactical communication with digital packets over a high-capacity common user radio (see Table 4.1).[6] By 1968, Welchman had advanced this concept into a communication architecture that he referred to as an *inverted U information pipeline* that could accommodate digital voice, digital image, and digital text data.[7] Welchman's architecture relied on a broadcast communications model rather than a point-to-point communications model.

TABLE 4.1

Major Events in Link 16 Development

Date	Event
Late 1960s	Gordon Welchman proposes inverted U information pipeline, which would be refined to become what is known as TDMA
1967	MITRE completes *Control and Surveillance of Friendly Forces* study, which proposes communication architecture based on Welchman's design
Late 1960s	Initial demonstration and Airborne Warning and Control System (AWACS) implementation
1976	JTIDS program office is stood up
1987	NATO buys Class 1 JTIDS terminals for NATO AWACS aircraft
1987	NATO endorses a Military Operational Requirement (MOR), formalizing a need for a common jam-resistant tactical communication link
1987	Multifunctional Information Distribution System (MIDS) program is initiated
1994	MIDS program passes Defense Acquisition Board review
1994	Link 16 is designated as DoD's primary tactical datalink for command, control, and intelligence (C2I) systems
2020	MIL-STD-6016 Revision G is published

---

[5] Northrop Grumman, 2014.

[6] Dyer and Dennis, 1998.

[7] This inverted U architecture would be refined to become what is known as time-division multiple access (TDMA). TDMA simply means that a communications channel is divided into time slots and time slots are assigned to different producers of data. In the case of Link 16, each JTIDS/MIDS unit on the network is allocated one or more time slots for their use. TDMA is contrasted most frequently with FDMA (or frequency division multiple access) in which each user is assigned its own frequency to communicate over, which it can use at any time. TDMA has two primary drawbacks: (1) All users must be time synchronized to avoid stepping on each other's communications and (2) unused bandwidth cannot easily be reallocated to other users. Link 16 uses what is called a dynamic TDMA communications paradigm.

Early development work to implement Welchman's ideas was executed by the MITRE Corporation.[8] In 1967, MITRE completed a study that concluded that battlefield information was being underused by the U.S. military. The study, *Control and Surveillance of Friendly Forces*, proposed a new communication architecture based on Welchman's ideas, whereby networked units would broadcast information to provide common situational awareness. This led to a MITRE-led development effort called Position Location Reporting and Control of Tactical Aircraft (PLRACTA). Using three ground-, two airborne-, and one truck-mounted terminals, PLRACTA successfully demonstrated the feasibility of the architecture. In 1973, PLRACTA was subsumed by Seek Bus, an Air Force–run project, which in, turn, was incorporated into the nascent AWACS program. Following a successful multination demonstration, it was merged with the U.S. Navy's JTIDS program in 1974. A joint service program office was appointed to administer the program.

By the late 1970s, Class 1 JTIDS terminals had been installed on AWACS aircraft.[9] Class 1 terminals were large (taking up an entire equipment rack on AWACS aircraft) and heavy (roughly 600 pounds). Because of their size and weight, they were only installed on the AWACS platform—derived from the Boeing 707—and in ground control facilities. The Class 2 JTIDS terminals were smaller and lighter and thus more widely deployed.

In 1987, NATO endorsed a MOR formalizing a need for a common jam-resistant tactical communication link, and the MIDS program was initiated to develop small form-factor lightweight Link 16 terminals on U.S. and NATO fighter aircraft.[10] In 1993, the MIDS program was restructured to employ an open architecture and incorporate commercial components into Link 16 terminals.[11] In October 1994, Link 16 was designated as DoD's primary TDL for C2I systems.[12]

The MIDS program office is developing the BU3 MIDS Joint Tactical Radio System (JTRS) terminal.[13] The BU3 configuration modernizes the electronics packaging using a radio frequency system-on-chip architecture. The terminal is expected to be complete by late 2021.

The most recent version of the Link 16 message standard is MIL-STD-6016, Revision G, which was published in 2020. The equivalent NATO standard is STANAG 5516, edition 8. Over time, the Link 16 message standard has expanded dramatically in scope as new platforms and capabilities have been incorporated into the standard. The current version of the Link 16 message standard is nearly twice as long as Revision B.

---

[8] Much of the information on the MITRE-led JTIDS development effort derives from a *Wikipedia* article written by C. Eric Ellingson, who was the head MITRE program manager on the original JTIDS program. A description of the JTIDS system is available online (see "Joint Tactical Information Distribution System," *Wikipedia*, last updated October 9, 2020).

[9] Myron Hura, Gary McLeod, Eric V. Larson, James Schneider, Daniel Gonzales, Daniel M. Norton, Jody Jacobs, Kevin M. O'Connell, William Little, Richard Mesic, and Lewis Jamison, *Interoperability: A Continuing Challenge in Coalition Air Operations*, Santa Monica, Calif.: RAND Corporation, MR-1235-AF, 2000.

[10] Daniel Gonzales, Daniel M. Norton, and Myron Hura, *Multifunctional Information Distribution System (MIDS) Program Case Study*, Santa Monica, Calif.: RAND Corporation, DB-292-AF, 2000.

[11] Gonzales et al., 2000.

[12] B. E. White, "Tactical Data Links, Air Traffic Management, and Software Programmable Radios," *Gateway to the New Millennium, 18th Digital Avionics Systems Conference, Proceedings*, Vol. 1, St. Louis, Mo., October 1999.

[13] Email correspondence, February 10, 2021.

Governance

MIL-STD-6016 Governance

Currently, Defense Information Systems Agency (DISA) is the DoD Executive Agent for joint interoperability of tactical command and control systems standards, including MIL-STD-6016. At the service level, four configuration boards are responsible for the administration of the standard: Army Interoperability Technical Review Board, Air Force Standards Interoperability Board, the Navy's Technical Interoperability Standardization Group, and the Marine's Technical Interoperability Standards Working Group. At the joint level, the configuration board is known as the Joint Multi-Tactical Data Link (JMT) configuration control board (CCB).

The process of making a change to MIL-STD-6016 typically begins with a new requirement within a Service branch, defense agency, or combatant command.[14] Once the service, agency, or combatant command has come to internal consensus about how to amend the standard to solve the requirement, they submit an interface change proposal (ICP) to DISA. The ICP contains information on the anticipated impact of the proposed changes on other military standards, the source of the requirement, and a markup with the proposed changes. At this point, other organizations may respond to the proposed changes with comments. Services review IPCs three times a year.[15] A majority vote by the JMT CCB, chaired by DISA, is required to approve the amended standard.[16] At this point, DISA marks the change with a CCB directive, which immediately allows any platform to implement the change. Because changes are incorporated into MIL-STD-6016 on a two-year cycle, the CCB directive allows implementation prior to the publication of an entirely new standard document.

The time needed to process a change depends on the size of the proposed change. Small changes, such as adding a new *store type,* which allows a user to label a new platform correctly, can be added quickly. In contrast, large changes that risk backward compatibility or require significant additions to the standard document, often take years to enter the standard.[17]

Link 16 Terminal Governance

The MIDS Program Office is responsible for the Link 16 waveform.[18] The MIDS program is a joint program involving the U.S. Army, U.S. Air Force, and the U.S. Navy. The U.S. Navy is the lead component of the program. The MIDS program also encompasses the MIDS International Program Office involving the United States, France, Germany, Italy, and Spain. The MIDS program is charged with administration of the production and acquisition of two products. The first is the MIDS Low-Volume Terminal, which is a lightweight JTIDS-compatible terminal that can transmit and receive MIL-STD-6016 messages. The second product is the MIDS JTRS, which is a software-defined radio terminal.[19]

---

[14] Authors' interview, February 2, 2021.

[15] Authors' interview, January 26, 2021.

[16] Once the Joint community has approved the amended standard, it will go forward for allied coordination: the process by which NATO approves proposed changes to the standard. NATO approval requires consensus by all voting member countries, but abstention is allowed.

[17] One such change was that to incorporate network-enabled weapons (NEWs). The changes to accommodate NEWs required hundreds of pages of changes to the standard. Authors' interview, February 2, 2021.

[18] Authors' interview, February 9, 2021.

[19] Gonzales et al., 2000; Authors' interview, February 9, 2021.

Link 16 Technical Characteristics

Link 16 employs a TDMA wireless network structure. The TDMA protocol divides time into discrete time slots (128 slots per second) to provide multiple communication channels. Because the time slots in the network are preassigned to the various participants, the network functions independent of the status of any individual participant.[20] This eliminates the need for a network control station, which can be a single point of failure in other tactical data links. Link 16 operates in the L-band frequency of the radio spectrum with aircraft using the 960–1,215 MHz portion of the band. Link 16 uses a frequency-hopping waveform that allows operation in a hostile electromagnetic setting. That is, Link 16 protects against narrow band jamming by using distinct transmission frequencies for each pulse to protect against interference from a frequency-band-matched jammer.

The MIL-STD-6016 messages are functionally oriented and use either fixed-word or variable message format.[21] The messages are identified using a label and a sublabel. [22] For example, the J.1.4 message is for connectivity status, which reports communication quality. In this identification scheme, the "J" indicates the message is a MIL-STD-6016 message, the "1" is the message label and "4" is the message sublabel.[23] Information exchange is limited to line-of-sight communication; however, using satellites and ad hoc protocols, it is possible to transmit MIL-STD-6016 messages over longer distances.

Link 16 and Warfighting Evolution

Today, Link 16 is used by dozens of U.S., NATO, and allied platforms, including the F-16, F-35, Joint STARS aircraft, Panavia Tornado, U.S. and allied nations carrier battle groups, THADD missile defense systems, various rotary wing aircraft, and unmanned aerial system (UAS) platforms. It is also used on mobile and fixed command bases.[24] The use of Link 16 is widely distributed across the Services. Table 4.2 provides a sample of systems across the U.S. Joint Force that use Link 16.[25] Link 16 is used for a wide spectrum of DoD missions, including close air support, defensive counterair, offensive counterair, air interdiction, combat search and rescue, theater command and control, command and control, air and missile defense, targeting and battle management.[26]

In the mid-1990s, the Air Force–run JTIDS Operational Special Project demonstrated the performance of Link 16 in tactical air-to-air combat. Operations of F-15Cs equipped with voice-only communications were compared with that of F-15Cs equipped with voice and Link 16 data communications.[27] Over the course of

---

[20] This is sometimes referred to in the TADIL J documentation as a *nodeless architecture*.

[21] Army Training and Doctrine Command, 2000.

[22] Northrop Grumman, 2014.

[23] Northrop Grumman, 2014.

[24] Missile Defense Advocacy Alliance, "Link 16," webpage, January 2017.

[25] The majority of the information contained in the table is derived from a TADIL J quick reference guide (see Army Training and Doctrine Command, 2000). We have added information on newer platforms and updated the information from older systems when appropriate.

[26] There was early skepticism about the utility of Link 16 from operators, especially among aviators. An engineer with more than 30 years' experience working with the Link 16 attributes early skepticism to negative experience with prior TDLs, especially Link 4. Authors' interview, February 9, 2021.

[27] Daniel Gonzales, John S. Hollywood, Gina Kingston, and David Signori, *Network-Centric Operations Case Study: Air-to-Air Combat With and Without Link 16*, Santa Monica, Calif.: RAND Corporation, MG-268-OSD, 2005.

TABLE 4.2
Select Platforms Using Link 16

Platform	Service
F-16 Fighting Falcon	U.S. Air Force
F-22 Raptor	U.S. Air Force
Air Operations Center	U.S. Air Force
E-3 AWACS B/C Sentry	U.S. Air Force
E-8 Joint STARS	U.S. Air Force
Air Defense Systems Integrator	U.S. Army
Patriot Information Coordination Center; AN/MSQ-116	U.S. Army
Forward Area Air Defense	U.S. Army
Advanced Combat Direction System	U.S. Navy
AEGIS, guided missile cruiser/guided missile destroyer (DDG)	U.S. Navy
Submarine (nuclear propulsion)	U.S. Navy
Marine Tactical Air Command Center	U.S. Marines
Tactical Air Operations Center AN/TYQ-23 (V1)	U.S. Marines

the exercise, more than 12,000 sorties were flown, and engagement conditions were varied with regard to quantity of combatants (ranging from two Blue fighters versus two Red fighters to eight Blue versus 16 Red) over different lighting conditions.[28] The primary independent variable of interest was the presence of Link 16 datalinks for the Blue forces. The experiment found that fighters equipped with the Link 16 datalink outperformed by a factor of 2.6 those equipped with only voice communications.[29]

A 2005 RAND study analyzing the outcomes observed in Link 16 equipped fighters found that the improved information environment enabled by the datalink led to faster and better pilot decisions, which led to improved performance. Interviews with experienced fighter pilots, including those who participated in the JTIDS Operational Special Project, found that Link 16 improved pilots' situational awareness and quality of information.

Over the course of the past two decades, Link 16 has had to evolve to meet the needs to today's platforms. Table 4.3 lists and briefly describes significant evolutions of the standard. In Chapter Five, we discuss the particular design and governance features that appear to have played a role in enabling these capabilities. Appendix B describes the capabilities in greater detail.

What Explains Link 16's Durability?

As described earlier, the intellectual origins of Link 16 trace back to at least 1968, when Welchman proposed a TDMA communication architecture.[30] As observed in the prior section, the standard remains in active use.

---

[28] Office of the Secretary of the Air Force for Acquisition, *JTIDS Operational Special Project (OSP) Report to Congress, Mission Area Director for Information Dominance*, Washington, D.C.: Headquarters U.S. Air Force, December 1997.

[29] Gonzales et al., 2005.

[30] Dyer and Dennis, 1998.

TABLE 4.3
Capabilities Enabled by Link 16

Capability	Role of Link 16 in Evolution or Advent
Improved tactical air-to-air combat	Link 16 improved information environment available to fighter pilots, which improved decisionmaking, leading to improved performance.
NEWs	MIL-STD-6016 was updated to allow precision-guided NEWs to accept in-flight targeting information. The changes to the messaging standard also allow the weapons to be handed off from one platform to another and thus to pursue more-mobile targets.
Battle damage assessment	The MIL-STD-6016 message set was extended to allow it to carry imagery from UASs over Link 16 to conduct battle damage assessment.
Communications beyond line of sight	In the late 1990s, the Missile Defense Agency (MDA) demonstrated the ability to take position information (i.e., tracks) from a Link 16 network, send it over satellite networks, and then display the data in distant theaters to cue additional sensors or shooters.
Connecting to DoD's Global Information Grid (GIG)	In 2005, Boeing granted a patent (US7953110-B1) for system performing tunneling of transmission control protocol/internet protocol (TCP/IP) packets over a Link 16 datalink, thus connecting fighter aircraft and other Link 16 terminal–equipped platforms to participate in the DoD's net-centric warfare concepts.
Interoperability with Army Battle Command Systems	Software was developed to transform messages between Army Battle Command Systems and Link 16.

Although the focus of this study is not on the determinants of the durability of standard, the longevity of Link 16 is notable and worth briefly exploring. Our research suggests that at least two variables have contributed to Link 16's longevity: its fulfillment of enduring and compelling operational needs and its proliferation across the U.S. joint and allied forces (i.e., its network effects).

Link 16 Fills Enduring and Compelling Operational Needs

Critical to the success of Link 16 is the role of the standard in fulfilling enduring operational demand. Our interviews consistently reveal that Link 16's role in providing near-real-time situational awareness, friend or foe identification, and clear digital voice communications explain Link 16's continued adoption.

Perhaps the most important capability from Link 16 is to provide platforms with a common tactical picture: a shared and identical representation of the battlefield that is updated in near real time. An engineer, embedded at MIDS Program Office, explicitly links the continued adoption of Link 16 to its role in providing situational awareness to operators, stating,

> There is a reason that platforms are continuing to implement it [Link 16] and that is that it works really well. It is a wholistic network in the sense that you have all these different platforms and all these different users, and they are all exchanging information, and everyone is refining their time off each other, their position off each other, you get broad situational awareness. [. . .] From a situational awareness point of view, Link 16 [is] phenomenal; it gives you access to the whole picture so to speak.[31]

---

[31] Authors' interview, February 1, 2021.

Link 16's secure digital voice communication capability also explains continued demand for Link 16 terminals. During our interviews, several engineers mentioned the clarity of Link 16 voice transmissions, especially compared with the traditional radios that Link 16 secure digital voice replaced. One engineer with extensive Link 16 experience described the audio quality of Link 16 voice as "crystal clear."[32] During another interview, a tactical data links expert described the advantages of Link 16 voice over traditional radios and also underscored the importance of Link 16's receipt acknowledgement:

> When I want to tell an aircraft to attack a target, if I'm telling him over a radio with [non Link 16] voice and the voice transmission is garbled, there is no data correction, there is no check sum, on a voice message. If I hear you say 'five' when you said 'nine,' which sound very similar on some kinds of radios, there can be a big difference on where I drop a bomb on which platform I attack. However, doing it on Link 16, the message is much faster, because the message is sent in seven milliseconds and there is also a message that goes back which is called a receipt compliance, which is exact copy of the message with only one byte shifted. . . . That allows the sender to make sure that the message was received properly, was understood, and was acknowledged, which can take seconds or minutes for humans to do over voice, but computers can in milliseconds. So, it [Link 16] removes a lot of those issues in communication.[33]

Link 16 also is valued for its role in mitigating the risk of fratricide by identifying friendly assets.[34] A senior tactical data link expert for the MIDS program describes the value of this capability based on his experience speaking to operators, stating, "I've spoken to a lot of operators who say that the ability of Link 16 to reduce Blue-on-Blue engagements is significant and it has helped them to secure the battlefield in [ways] that cannot be quantified."[35]

---

[32] Authors' interview, February 9, 2021.

[33] Authors' interview, January 26, 2021.

[34] Note that Link 16 itself does not identify friend versus foe, but it is designed to communicate the results of such decisions that are made both to a given weapons system and across weapons systems.

[35] Authors' interview, January 26, 2021.

CHAPTER FIVE

Standards and Warfighting Evolution

Chapters Four and Five identified evolving military capabilities that were enabled, at least in part, by STANAG 4607 and the Link 16 message standard. This section discusses features of the standards, and their associated ecosystems, that facilitated the observed warfighting evolution. We identified three features that appear to play an important role in facilitating capability evolution: built-in extensibility, the incorporation of operational feedback, and an active user community with data access. We also discuss a negative effect of standardization: Although Link 16 might have increased the potential for interoperability across the kill chain, the proliferation of distinct implementations might harm interoperability. Finally, we discuss how Link 16's failure to adhere to a layered communication model has limited future warfighting evolution and the two most-common methods of dealing with this limitation: tunneling and translation.

Built-in Extensibility

Both STANAG 4607 and the MIL-STD-6016 were designed to be extensible. Both standards included a substantial number of unpopulated message segments that were intended to allow the subsequent inclusion of new information. Further, both standards have formal, transparent, and relatively open processes for extending the standard.

Built-in Extensibility in STANAG 4607

As described earlier, designers devised a means of extending the STANAG 4607 format to incorporate data from space-based sensors and multichannel SAR-based GMTI sensors.[1] Extensions made to accommodate SAR sensors include adding segments for the measurements of target slant range and cross-range positions.[2] One of those designers is explicit in citing the extensibility of the STANAG 4607 as a motivation for using it as the baseline for the RADARSAT-2 GMTI Technology Demonstration Project.[3] Similarly, when asked about STANAG 4607's ability to enable warfighting evolution, a subject-matter expert we interviewed stated, "The ability to add segments or new types of segments to the data allowed us to be flexible."[4]

---

[1]   Pierre D. Beaulne, *The Addition of Enhanced Capabilities to NATO GMTIF STANAG 4607 to Support RADARSAT-2 GMTI Data*, Ottawa, Canada: Defense Research and Development, technical memorandum, DRDC-O-TM-2007-341, December 2007; Rahim Jassemi-Zargani and Pierre D. Beaulne, "SBR MTI Impact on Coastal and Border Surveillance," meeting proceedings, NATO Standardization Agency, Science and Technology Organizations, RTO-MP-SET-125-21, April 18, 2008; Authors' interview, 2021.

[2]   Jassemi-Zargani and Beaulne, 2008.

[3]   Beaulne, 2007. The direct quote is: "Since STANAG 4607 is expandable, it was decided that it should be modified to support spaceborne SAR-GMTI platforms" (see Beaulne, 2007).

[4]   Authors' interview, January 20, 2021.

Evidence of the importance of built-in extensibility also comes from the *process* by which the STANAG 4607 CST modified the standard to accommodate data from space-based platforms. As described above, the advent of new SAR radars on NATO member reconnaissance satellites created demand to transmit GMTI data from space. Faced with this demand, the CST faced three options: (1) use the standard as is, (2) modify the existing data segments by making spaceborne sensor-specific exceptions or caveats, or (3) add additional space-specific data segments.[5]

Our interviewees explained that the first option—attempting to use an unmodified version of the STANAG 4607 to transmit spaceborne GMTI data—was a nonstarter given the distinct demands of space-based sensors.[6] The second option was possible but had several drawbacks. For example, the CST could have modified the standard to include sensor-specific metadata for existing data segments and then tasked recipients with interpreting the data based on externally derived knowledge of the sensor type. This was judged as unnecessarily complicating the interpretation task for the recipient. Another way to implement the second option would have been to transmit the space-specific information via STANAG 4607's existing free text data segment. However, this was judged to unnecessarily complicate backward compatibility.[7] As one interviewee noted, "We [the CST] could have put our space-based data into the free text, but then you are really violating the intent and you sacrifice compatibility with the community."[8] Ultimately, the CST selected the third option: using STANAG 4607's built-in extensibility to add space-specific data segments. As an interviewee noted, "We decided let's just add additional data segments that don't in any way perturb how the airborne community is already using it; let's just add a couple of space-specific data segments that allow us to do what we need to do in the space world."[9]

The method—adding the new message segments without modifying existing segments—was enabled by the STANAG 4607's inclusion of a process, and hooks, to add new message segments.[10] Besides being deliberate, STANAG 4607's extensibility is transparent. The standard, and its implementation guide, describe how to include new capabilities, such as radar modes, platform types, or data processing techniques, while maintaining forward and backward compatibility of the standard.[11] The process to make a change to the standard is defined in the STANAG 4607 Implementation Guide.[12] This process, open to any national representative, is straightforward and is initiated via submission of a change request form to the CST.

In addition to providing additional segments for unknown future capabilities, the original STANAG 4607 developers also left room for known (albeit, at that point, immature) capabilities. For example, the HRR message segment was one of the original STANAG 4607 (edition 1) message segments,[13] and the NATO GMTI

---

[5] Authors' interview, January 8, 2021.

[6] Authors' interview, January 8, 2021.

[7] Authors' interview, January 8, 2021.

[8] Authors' interview, January 8, 2021. "It is not intended to be a means by which developers are able to circumvent the segments provided within STANAG 4607" (AEDP-7, 2013, p. I-7). The interviewee also noted that this would be "a duct tape solution."

[9] Authors' interview, January 8, 2021.

[10] As one interviewee noted, "Fortunately, the way the STANAG was laid out, it permitted it to be a living structure so that we could add additional data segments as necessary." Authors' interview, January 8, 2021.

[11] Beaulne, 2007.

[12] AEDP-7, 2013, p. M-2.

[13] The fact that STANAG 4607 was able to support HRR chips, a capability that requires high bandwidth and is not available in older exploitation systems, also owes to the inherent scalability of the standard. As described above, the format divides data into segments. These segments are permitted to be transmitted in any order and not all segments must be transmitted. This allows a system that is using the standard to only receive fields that it needs or is capable of processing. The scalability

TST established in April 2000 chartered the HRR working group to focus on this area. The inclusion of an HRR working group when only a small subset of exploitation systems had the bandwidth necessary for HRR transmission illustrates the forward-looking approach taken by STANAG 4607's original developers.

STANAG 4607's forward-looking design appears to have been at least partially driven by operational considerations. More than one of our interviewees stated that operational demand to use GMTI data for time-sensitive targeting informed early design decisions.[14]

Built-in Extensibility in Link 16

The MIL-STD-6016 message standard is also extensible and has well-defined processes for adding features. It includes unspecified message types that can be defined as needed for future use. Our interviewees stated that the inclusion of unspecified message types was an intentional means to accommodate future demand from new platforms and technologies.[15]

Additionally, Link 16 has a message segment, J28 messages or National Use Message Segments, that is excepted from some of the standard's interoperability requirements. This exception allows for experimentation and prototyping of features prior to their inclusion in the formal standard. For example, this segment was used by Air Force engineers to demonstrate integrated fire control prior to incorporation of that capability into the formal standard.[16]

The Incorporation of Operational Feedback

Important changes to the standards themselves (e.g., changes that support inclusion of new sensor types and the removal of ambiguity in message standard implementation guidance) have been driven by feedback from real-world operations.

Incorporation of Operational Feedback in STANAG 4607

In the case of STANAG 4607, demand from operators motivated the initial establishment of an open GMTI standard. As described in Chapter Three, during the 1990–1991 Gulf War, Joint STARS used a proprietary GMTI standard.[17] This proprietary standard filled a narrow GMTI requirement: the low-bandwidth transmission of GMTI data to specific Army CGS platforms. However, operators during and following the conflict sought broader use of GMTI data. This spurred demand for an open GMTI military standard.

Demand from operators also played a role in determining the initial technical content of STANAG 4607. The 1990–1991 Gulf War operators saw the need for GMTI data to conduct time-sensitive targeting of moving

---

of transmissions allows the inclusion of sophisticated capabilities, such as the transmission of HRR chips, without sacrificing backward compatibility—the condition that older systems be able to use future versions of the standard.

[14] A direct quote from one of the interviewee's reads,

> So the people who eventually got involved in working on 4607 were aware of the demand for time-sensitive targeting and said hey there are lot of things that can be done with these radars in the future, not just what was originally thought of as the Joint STARS requirement [low bandwidth transmission of GMTI data to Army exploitation systems]. (Authors' interview, January 4, 2021)

[15] Authors' interview, February 3, 2021.

[16] Authors' interview, February 3, 2021.

[17] This standard was proprietary to the original developer, Northrop Grumman. Authors' interview, January 4, 2021.

ground vehicles, such as the TEL vehicles used to transport Iraqi Scud missiles. This demand, often articulated within the STANAG 4607 community of interest, contributed to the inclusion of an HRR message segment that became a critical enabler of warfighting evolution. As noted by one of our interviewees, "the whole concept of high range resolution was an important piece of [time-sensitive targeting and target identification] because it was a way to discriminate targets."[18]

The Incorporation of Operational Feedback in Link 16

The advent of Link 16 is owed largely to operational demand. Welchman's proposed proto-TDMA architecture, and MITRE's initial implementation, were driven by the perception of underutilization of battlefield information by the U.S. military. Furthermore, the incorporation of operational feedback was critical in correcting interoperability failures that developed because of service-specific approaches to implementing Link 16 during the 1980s. These differences in how platforms and services implemented Link 16 might have been the result of, and perhaps were directly facilitated by, ambiguity in the text of MIL-STD-6016. There is anecdotal evidence that, in certain cases, the standard was written to accommodate distinct service-specific implementations. During an interview for this report, an engineer who worked on MIL-STD-6016 during this period described how distinct implementations by the U.S. Army, the U.S. Air Force, and the U.S. Navy affected how the standard was written, stating,

> What ended up happening in the '80s is that we had one standard with three different ways of doing things. There was almost a tacit agreement that if there was a difference in opinion, the standard would be written ambiguous enough that everybody could interpret it to their own liking.[19]

These differences in Link 16 implementation led to interoperability challenges during conflict. During the 1990–1991 Gulf War, some Link 16–equipped platforms did not receive all expected Link 16 transmissions. This led to a condition where "AWACS [an AF platform] could control Air Force fighters but couldn't control the [Navy] F-14, which had Link 16 at the time. And the Navy couldn't control Air Force fighters, but they could control their F-14s."[20] The root cause was divergence in how the U.S. Air Force and U.S. Navy implemented the standard protocol for an aircraft to check in to the network. For the U.S. Air Force, the check-in was made optional and for the Navy, the check-in was mandatory. In practice, however, U.S. Air Force platforms simply did not implement the check-in and therefore could not interoperate with U.S. Navy control systems that expected them to check in.

Eventually, MIL-STD-6016 was modified to remove the ambiguities that led to the interoperability failures observed during the 1990–1991 Gulf War. An engineer who worked on Link 16 for over thirty years points out that lack of realism in training scenarios contributed to why these issues were not detected sooner, stating, "In developmental testing most of the tests are scripted for success, but in real live operations almost nobody follows a script and that is where you find your real problems [. . .] It was only the point at which we deployed for real operations that it [the standard] dramatically improved."[21]

Another example where operational demand drove innovation is Link 16's capability to transmit images from a UAS in support of battle damage assessment while reducing pilot risk. A TDL expert within the MIDS

---

[18] *Discrimination of targets* is the ability to distinguish the true target (e.g., a warhead) from other objects (e.g., chaff or other decoys designed to confuse targeting systems). Authors' interview, January 4, 2021.

[19] Authors' interview, February 3, 2021.

[20] Authors' interview, February 3, 2021.

[21] Authors' interview, February 3, 2021.

program explained the rationale for the development of the capability as follows: "It is risky to send human eyeballs into a combat zone just to do battle damage assessment. So, one thing they realized was they could send a drone to take a picture and send the picture back to the command center."[22]

User Community and Data Repository

The STANAG 4607 user community and the Atlantis GMTI data repository also appear to have driven significant warfighting evolution.

By providing a publicly available, well-specified means of interacting with GMTI data, STANAG 4607 enabled the development of a wide range of third-party applications. As a former engineer at MIT Lincoln Laboratory explained,

> When we got data off Joint STARS, it would be in this standard format [STANAG 4607] and we knew that we had parsers that we could use, we could run our MATLAB scripts and we could replace all sorts of tracking algorithms because we knew what the [radar] detections were going to look like. I always found it very helpful.[23]

As described above, during the mid-1990s, a GMTI community of interest was established. The effect of the community was to establish an active base of GMTI data users that would develop algorithms to address operational needs. For example, the STANAG 4607 user community played a role in the development of applications using GMTI data to counter-IED operations by forming a community of interest with regularly scheduled meetings. This venue, attended by operators, program staff, and developers, facilitated the exchange of information about how to address real-world challenges. This vibrant STANAG 4607 user community was enabled by the standard's openness. Such third-party collaboration is less likely with closed, or proprietary, standards. The vibrancy of the STANAG 4607 user community was also enabled by the creation and sharing of common software libraries and tools. AFRL provided the community with JAVA, C++, Matlab code and tools to read, parse, filter, modify, stream, record, validate, and convert STANAG 4607 data.[24]

Also critical to the warfighting evolution enabled by STANAG 4607 was the creation of the Atlantis GMTI data repository in 2003. This repository provided direct access to more than 15 years' worth of STANAG 4607–formatted data for third-party algorithm development and testing.[25] Notably, the data repository was built with capability enhancement in mind. As explained to us by a senior engineer who used the repository,

> [T]o have a one-stop shop for GMTI data where you'd have a mix of platforms, areas, environments (rural, urban, and surface) would allow us to explore the algorithms we were developing. [. . .] That allowed us to more efficiently build tracking and fusion algorithms that were highly effective with current generation sensors.[26]

The Atlantis GMTI–stored data allowed users to make pattern-of-life observations of adversary movements over time with algorithms developed to do group tracking, forensic backtracking, meeting detection, route avoidance, anomaly detection, and road and traffic detection. This allowed for the extraction of intel-

---

[22] Authors' interview, January 26, 2021.

[23] Authors' interview, December 23, 2020.

[24] Email correspondence, February 18, 2021.

[25] Authors' interview, January 4, 2021.

[26] Authors' interview, January 20, 2021.

ligence from the GMTI data, expanding the use of radar sensors far beyond that of a real time command and control resource).[27] As one interviewee observed, "AFRL built the database, provided it to SIPRNet users, and that is where the story begins in terms of forensic backtracking, IED emplacement, and counter-IED work. Certainly, if all that data was a different format per platform, it would have been a challenge."[28]

One of the means by which the Atlantis GMTI data repository drove capability development was by extending the active GMTI user base, which currently totals around 3,300. One observer noted, "There was a whole cottage industry that grew up around the idea of what kind of information can we get out of this GMTI data."[29] Today, analysis and development cells at MITRE, the National Air and Space Intelligence Center (NASIC), NGA, and Hanscom Air Force Base continue to use STANAG 4607–formatted GMTI data from the Atlantis GMTI data repository in their work.

Brevity and Scope of the Message Standards: Can the Proliferation of Distinct Implementations Harm Interoperability?

Our research does not allow us to reach a firm conclusion about the role a standard's scope plays in facilitating warfighting evolution. Our research in this regard is ambivalent. Although some evidence suggests that Link 16's broad scope facilitates interoperability, other evidence suggests that wider scope correlates with a profusion of integration heterogeneity. Here we briefly present both arguments.

Including appendixes, the current STANAG 4607 standard (edition 3) is 107 pages. The STANAG 4607 implementation guide, including appendixes, is only 301 pages. In contrast, the current version of MIL-STD-6016 is more than 14,000 pages.[30]

MIL-STD-6016 is used by platforms ranging from ballistic missiles to DDGs to ground stations, across all Service branches and by many allied nations. Its functional scope is wide, including control of fighter aircraft, exchange of compressing digitized voice, and encryption.[31] On one hand, the standard's scope may increase interoperability by facilitating the exchange of information across a given kill chain, including across domains, platforms, and services. When asked about the role of the standard's scope on warfighting evolution, one of our interviewees described this possibility thusly:

> The size of the standard allows for a diverse set of messages to enable ground-, sea-, and space-based information exchange. Consider the Ballistic Missile Defense System (BMDS): Link 16 allows for the exchange of space track (J3.6), weapons inventory (J13), and similar information spanning Army, Air Force, Navy, and other U.S. and Allied assets. This data is used across the kill chain—from enemy threat launch to threat intercept—to cue radars, provide situational awareness to Warfighters, correlate tracks for dissimilar sources, etc. Within this context, Link 16 has been successfully demonstrated within multiple Missile Defense Agency (MDA) ground and flight test events of the BMDS.[32]

On the other hand, we also observed negative effects of MIL-STD-6016's wide scope. One effect stems from the proliferation of distinct implementations of the message standard. Although all Link 16–enabled

---

[27] Authors' interview, January 20, 2021.

[28] Authors' interview, January 20, 2021.

[29] Authors' interview, January 4, 2021.

[30] It is worth noting that, when a party is implementing the standard, only a small subset of its messages and content are used.

[31] Authors' interview, February 3, 2021.

[32] Email correspondence, February 20, 2021.

systems are required to be compliant with the standard, the autonomy given to the military services in implementing the standard could lead to distinct versions in service at any given time. As one experienced engineer noted,

> Control belongs to the services [. . .] If the Navy wishes to adopt MIL-STD-6016 Foxtrot and the AF wants to go to Golf [i.e., the F vs. G revision of the standard], that is their choice. And whatever differences exist between those standards, they are going to have to live with.[33]

The proliferation of implementations also drives a need to extend the message standard to accommodate differences because, once a platform has implemented a given approach, that platform, in essence, forms a constituency against further change. This inflexibility could drive a stubborn complexity in the standard, described to us as,

> Link 16 has become hard to manage because with Link 16 and most standards, you can never go back and fix things. Once there is a rule, there is going to be somebody that implemented it, and they are going to argue against it being modified. So, there are a lot of rules that are still there, even though they were proposed in the mid 70s and that in my opinion are obsolete and should be eliminated.[34]

Impacts of Lack of Conformance to a Layered Communication Architecture

The age of Link 16 imposes limits on its compatibility with more-modern approaches to communication system designs, such as those used in the standard four-layer internet model.[35] Applying a layering approach to Link 16 poses difficulties because it is "a highly integrated stovepipe system with a high level of interaction between the different (potential) layers; and certain layering inconsistencies have been found. As a result, it would be difficult to incorporate all of Link 16 into the Global Grid."[36]

A 1999 paper considered the application of the ISO seven-layer model to Link 16 and observes that, within a layering approach, functions are meant to be portioned such that "each self-contained physical entity within a system realizes only functions within the same layer of the architecture."[37] However, Link 16 violates this principle. The paper notes that parts of the Link 16 cryptographic management functions are located within the message signal processor. During an interview for this report, an engineer with experience working with the standard described Link 16 at that time as "a pile of spaghetti. There was one thing connected to another thing [. . .] If you change something in Link 16 on one area, you've got to change it in so many places."[38]

However, the influence of a nonlayered approach has changed over time. As we described in Chapter Four, engineers no longer try to detangle Link 16 but have instead adopted tunneling and translation techniques when adapting it to more-modern networks.

---

[33] Authors' interview, January 26, 2021.

[34] Authors' interview, February 3, 2021.

[35] Warren J. Wilson, "Applying Layering Principles to Legacy Systems: Link 16 as a Case Study," *2001 MILCOM Proceedings Communications for Network-Centric Operations: Creating the Information Force*, Vol. 1, McLean, Va., October 2001.

[36] Wilson, 2001, p. 531.

[37] White, 1999.

[38] Authors' interview, January 19, 2021.

Conclusions: Does Design Matter?

From our examination of these two standards, we found that design does not appear to be a primary determinant of capability evolution. The techniques used to allow the standard's extension via extra messages are very simple in both design and implementation. Both standards enabled significant warfighting evolution and, in the case of Link 16, this was almost despite the inelegance of its design. These cases demonstrate that the network effects of an open and implemented standard (even in the cases where it was poorly implemented) can far outweigh any design considerations. These network effects were enabled by good governance, active users, and a strong tie to the operational warfighting community.

The STANAG 4607 community found success by transforming from a command and control message exchange standard to a data-centric repository that can be used by ever-more-capable algorithms to provide operational value. That success foreshadows DoD's recent initiatives to become a more data-centric organization. At some level, its success might lie in the narrowness of its focus. By concentrating only on a single type of sensor (radar) and a single type of information (ground moving targets), the community may have been better positioned to make the changes necessary to transition from a message-centric to a data-centric concept of use. The use of a formal data model as part of its specification also undoubtedly helped.

In contrast, the monolithic architecture and wide scope of Link 16 has limited its use in more-modern architectures, even as a command and control system. Furthermore, because it was designed as a digital substitute for voice communications between human pilots, the information conveyed across its links is not well formatted for machine-to-machine communication. Yet, despite these obvious drawbacks, it persists in large part because of its large installed base. Switching costs are extremely high. Link 16's success may be a hindrance to future evolution, reminding us that although good design may not be essential to short-term evolution, poor design can lock us into current solutions that make it impossible to evolve further. The design principles of layering and the separation of concerns developed in the years since Link 16 was first designed will, if adhered to, ensure that future DoD standards can achieve the network effects of standardization while mitigating switching costs.

Our findings regarding extensibility and governance lead to our suggested guidance on standard design and governance. To enable interoperability and capability evolution, we suggest that standards and their governing bodies should

- be designed with a built-in technical means of extensibility
- lay out an open and transparent means of drafting and amending the standard
- seek feedback from operators at all stages of the standard's life cycle
- cultivate a user and supplier community that allows these individuals to provide feedback into the standard design and amendment process.

Regarding design paradigms of the standards themselves, we found that the standardization of data, which occurred in the case of STANAG 4607, yields evolution and innovation, while the standardization of transport and link layers, which occurred in Link 16, has inhibited innovation. We believe this is because data are a resource that can be exploited, while transports and links are constraints that must be overcome. Our design guidance for consideration thus mirrors two of the data decrees from a DoD official's May 2021 memorandum regarding the creation of data advantage:

- Treat data as assets, which would include the creation of common interface specifications.

- At the transport and link layer, "use automated data interfaces that are externally accessible and machine-readable; [. . .] use industry-standard, non-proprietary, preferably open-source technologies, protocols, and payloads."[39]

---

[39] Kathleen Hicks, "Creating Data Advantage," memorandum from the Deputy Secretary of Defense to Senior Pentagon Leadership Commanders of the Combatant Commands Defense Agency and U.S. Department of Defense Field Activity Directors, Washington, D.C., May 5, 2021.

Description of Capabilities Enabled by STANAG 4607

Supportability of HRR Chips

Many modern radars have HRR mode, which allows a radar to attain a detailed representation of a target.[1] These representations can be stitched together to produce what is known as an HRR chip, a high-resolution image of a target that facilitates target identification. The STANAG 4607 standard today supports HRR chips through the HRR message segment.

As pointed out by one of our interviewees who had used STANAG 4607–formatted data for Defense Advanced Research Projects Agency's Vehicle and Dismount Exploitation Radar program, when the standard was first drafted, HRR chips were not yet common. Regarding the utility of this capability, this individual noted,

> The STANAG format was able to capture those [HRR chips], so that when we generated a detection, we could store a couple of chips of what the thing looked like, so that someone else down the line could look at those chips and see if two contacts were the same thing. So, STANAG was really flexible. When STANAG was first envisioned, I can't imagine they were using high-range resolution chips.[2]

Use of Space-Based Sensors for GMTI

Although initially designed for airborne GMTI sensors, the STANAG 4607 format has evolved to accommodate data collected by military and commercial satellites. The extension of STANAG 4607 to include space-based systems occurred soon after the promulgation of edition 1 of the standard and was driven by a technology push. During this period (roughly 2005 to 2007), several NATO member satellite systems with the potential capability to transmit GMTI data were coming online. Examples include Canada's RADARSAT-2, Germany's SAR-Lupe and TerraSAR-X, and the U.S. space-based radar program. Some sources state that the RADARSAT-2 program was the major impetus for this evolution of the standard.[3] Changes were made in edition 2 of the standard to enable transmission of data from these spaceborne satellites.

Edition 1 of the standard used a geodetic system to communicate the position of the platform on which the radar is mounted. However, using a geodetic system to communicate the position of a space-based plat-

---

[1] The range resolution of a radar is the minimum separation (in range) of two targets of equal cross section that can be resolved as separate targets. Sensors that can resolve smaller separations are referred to as HRR sensors.

[2] Authors' interview, December 24, 2020.

[3] The effort to extend the standard to accommodate GMTI data from new reconnaissance satellites is described in Jassemi-Zargani and Beaulne, 2008.

form results in a loss in accuracy. Therefore, the standard was extended to allow position information to be expressed in an Earth-centered Cartesian coordinate system. Additionally, because the look angle from a spaceborne platform significantly affects the imagery it collects, a new segment was added to record the sensing platform's orientation.[4]

Extending the standard to accommodate space-based sensors is expected to create new operational capabilities. For example, using GMTI data from space-based platforms might enable more-effective coastal surveillance and border security operations.[5] A 2008 paper includes results from modeling and simulation demonstrating that adding space-based GMTI data to that available from existing airborne ISR assets increases the surveillance coverage area resulting in earlier target identification. More recently, the U.S. Space Force announced that it plans to take on the task of providing GMTI data from space sensors for those cases when aircraft-based systems are operating in highly contested environments and thus may not be able to collect data.[6]

Extending Application to Multichannel SAR

SAR radar is often used in military applications because of its ability to render high-resolution images from an airborne or spaceborne platform largely independent of flight altitude and weather conditions. The early edition of STANAG 4607 could not accommodate SAR data. For edition two of the standard, it was expanded to accommodate data collected by these sensors. Extensions include adding segments for the measurements of target slant range and cross-range position.[7]

Using this development effort, the current version of STANAG 4607 can accommodate the multichannel SAR information from space-based sensors, such as the Canadian Space Agency's RADARSAT-2 platform. Additionally, the standard can accommodate *inter alia*, the APY-8 (a SAR radar made for UASs), and the ASARS-2A (another SAR radar).

Use of GMTI Data to Track IED Attacks

In the early 2000s, U.S. troops in Iraq and Afghanistan faced the onset of a new threat: IEDs. In efforts to counter the IED threat, the U.S. military sought to exploit all possible data sources, including GMTI data. Specifically, it was hoped that once an IED detonation was detected, historical GMTI data fused with data from other sensors could be used to track the individuals who had placed the explosive.[8] To this end, analysis and development cells in MITRE, the NASIC, NGA, and Hanscom Air Force Base used STANAG 4607–formatted GMTI data to develop counter-IED applications.[9]

---

[4] The orientation of a space platform is measured as its roll, pitch, and yaw about the x, y, and z axis of a coordinate system fixed to the body of the platform. This is analogous to aircraft roll, pitch, and yaw where the x axis is aligned to the orbital path of the satellite.

[5] Jassemi-Zargani and Beaulne, 2008.

[6] Tara Copp, "Space Force Aims to Take on an Air Force Surveillance Mission," *Defense One*, May 12, 2021.

[7] These efforts are described in Beaulne, 2007, and Jassemi-Zargani and Beaulne, 2008. Beaulne was part of the technical support team responsible for extending the STANAG 4607 format to accommodate space-based sensors and SAR-GMTI sensors.

[8] Authors' interview, January 4, 2021.

[9] Authors' interviews, January 4, 2021, and January 20, 2021.

Use of GMTI for Intelligence Purposes

STANAG 4607 enabled greater use of GMTI data for intelligence purposes. Although the traditional use of GMTI data is for real-time command and control functions, more recently it has been used to produce intelligence from stored data. Through the development of novel algorithms, historical GMTI data can be fused with other sensor data to yield valuable intelligence. One interviewee noted that AFRL, NASIC, NGA, and the Air Force Life Cycle Management Center all use STANAG-formatted data from the Atlantis data repository to develop algorithms for intelligence purposes. These algorithms include group tracking, forensic backtracking, meeting detection, route avoidance, anomaly detection, road detection, and traffic detection.[10]

---

[10] Authors' interview, January 20, 2021.

Description of Capabilities Enabled by Link 16

Network-Enabled Weapons

NEWs are precision-guided munitions that can receive updated targeting data or an abort message in flight from connected platforms. Additionally, the platforms can hand off the weapon to provide visibility beyond the line of sight of the platform that launched the munition. This hand-off capability provides the United States and our allies with the ability to maintain positive control of precision weapons even when engaging highly mobile targets. The MIL-STD-6016 NEW message set supports this significant enhancement to warfighting capabilities.

Battle Damage Assessment

Another warfighting evolution enabled by Link 16 is UAS-enabled battle damage assessment. Specifically, the MIL-STD-6016 message set was extended to allow UAS-collected imagery to be transmitted back to remote operators via Link 16. This capability to perform a battle damage assessment using remotely operated UAS allows commanders to avoid putting human operators in harm's way.[1]

Communications Beyond Line of Sight

The use of L-band frequencies limit Link 16 transmission range to line-of-sight communications unless aided by relays. The most common means of increasing the range of Link 16 beyond the line of sight is by using other airborne assets as relay points.[2] However, using the Link 16 waveform as a relay is quite inefficient because it can double the number of time slots needed for communications. Instead, satellites can be used to extend the range of Link 16 communications beyond the line of sight, but this requires separating the MIL-STD-6016 messages from the underlying waveform.

Davis et al., 1997, describe a development and demonstration project to extend Link 16 information transmission range beyond line of sight using both ultra-high-frequency and extremely high-frequency military communication satellites.[3] This TADIL J Range Extension (JRE) demonstration program, run by the Ballistic Missile Defense Organization (now the Missile Defense Agency), succeeded in extracting the MIL-STD-6016

---

[1] Authors' interview, January 26, 2021.

[2] Northrop Grumman, 2014.

[3] Barry W. Davis, Cecil Graham, David Stamm, and Chris Parker, "Tactical Digital Information Link (TADIL) J Range Extension (JRE)," *MILCOM 97 Proceedings*, Vol. 1, Monterey, Calif.: IEEE and AFCEA, November 1997, pp. 408–412.

track messages sent by theater air defense platforms, filtering the messages to meet the satellite data rate requirements, sending them over the satellite links, and reassembling them on receipt to the standard format for display on JTIDS devices.[4] The demonstration culminated during the 1996 All Service Combat Identification Evaluation Test in Gulfport, Mississippi. During this event, the JRE passed Link 16 messages beyond the line of sight from a *Ticonderoga*-class cruiser located in the Gulf of Mexico to an air defense artillery operations center in Gulfport using ultra-high-frequency satellite terminals.

This demonstration did not require changes to either the MIL-STD-6016 or to the Link 16 terminals. Instead, it was implemented with the JRE as a standalone system that exchanged MIL-STD-6016 messages from a Link 16 terminal over a 1553 data bus. Davis reports that the prospect of embedding the satellite link capability directly into a Link 16 terminal has many disadvantages, and so the standalone implementation was used. Despite the inelegance of the implementation, the utility of being able to send track messages beyond line of sight to cue other sensors was judged to be of great value. Today, the MDA has implemented translators to many different track message formats (including MIL-STD-6016) to provide integrated missile defense.

Connecting to DoD's GIG

The Air Force funded a development effort to improve the ability of Link 16 to interoperate with other components of the DoD's emerging GIG and the internet.[5] As a result of the effort, Boeing developed a means (US7953110B1) to allow tunneling of TCP/IP packets over the Link 16 datalink. This allows any application capable of outputting TCP/IP-based communications to use the Link 16 waveform as a link layer and thus benefit from its secure, anti-jam capabilities. As with the upgrades to use satellites as the underlying network layers for MIL-STD-6016 messages, this evolution is only made possible by a rather inelegant implementation—in this case, tunneling.[6]

Interoperability with Army Battle Command Systems

Elledge et al., 2007, developed a means to improve battlefield situational awareness for the U.S. Army aviation community by creating a software that enables exchange of information between the Link 16 network and Army aviators via the Army Battle Command Systems.[7] The software transforms XML messages (using an eXtensible Stylesheet Language Transformation) to enable the transmission of Link 16 information, via Army Battle Command Systems, to Army helicopter platforms.[8] By allowing U.S. Army aviation platforms to transmit and receive information (e.g., positional) to and from Air Force platforms, the software improved

---

[4] A MIL-STD-6016 track message contains the identity (friend, foe, neutral), classification (commercial aircraft, fighter jet, etc.) and position of a sensed object on the battlefield.

[5]   Steven A. Dorris, David E. Corman, Thomas S. Herm, and Eric Martens, "TCP/IP Tunneling Protocol for Link 16," U.S. Patent 7953110B1, filed January 11, 2010, assigned May 31, 2011.

[6]   To tunnel through a network is to ship a foreign protocol across a network that normally would not support it. This inevitably creates inefficiencies because the foreign protocol (including all of its header information) must be encapsulated within the data element of the native packet. In this case, the internet protocol packets are encapsulated within a MIL-STD-6016 message segment.

[7]   Ashley Elledge, Terrance Zimmerman, Bruce Robinson, and Jasen Jacobsen, "Exchanging Link 16 CoT Messages and US Army ABCS Messages via PASS," *MILCOM 2007, IEEE Military Communications Conference*, Orlando, Fla., October 2007.

[8]   Authors' interview, February 5, 2021.

situational awareness for joint operations. Furthermore, the proposed solution uses existing hardware on both sides of the transmissions, allowing "for the exchange of information without adding equipment to army aviation platforms."[9] Again, this is an example of an upgrade to the underlying command and control system without affecting the waveform or the message standard.

---

[9] Elledge et al., 2007, p. 1.

Abbreviations

AEDP-7	Allied Engineering Documentation Publication Number 7
AFRL	Air Force Research Laboratory
ASD/C3I	Assistant Secretary of Defense for Command, Control, Communications and Intelligence
AST	administrative support team
AWACS	Airborne Warning and Control System
C2I	command, control, and intelligence
C4ISR	command, control, communications, computers and intelligence, surveillance, and reconnaissance
CCB	configuration control board
CGMTI	common ground moving target indicator
CGS	Common Ground Station
CST	custodial support team
DDG	guided missile destroyer
DII COE	Defense Information Infrastructure Common Operating Environment
DISA	Defense Information Systems Agency
DoD	U.S. Department of Defense
DSK	Dvorak Simplified Keyboard
FNC3	fully networked command, control, and communications
GIG	Global Information Grid
GMTI	ground moving target indicator
HRR	high-range resolution
ICP	interface change proposal
IED	improvised explosive device
IETF	Internet Engineering Task Force
IPT	integrated project team
IPv4	internet protocol version 4
IPv6	internet protocol version 6

ISO	International Organization for Standardization
ISR	intelligence, surveillance, and reconnaissance
JMT	Joint Multi-Tactical Data Link
JRE	J Range Extension
JTIDS	Joint Tactical Information Distribution System
JTRS	Joint Tactical Radio System
MDA	Missile Defense Agency
MIDS	Multifunctional Information Distribution System
MIL-STD	Military Standard
MOR	Military Operational Requirement
NASIC	National Air and Space Intelligence Center
NATO	North Atlantic Treaty Organization
NDP	Neighbor Discovery Protocol
NEW	network-enabled weapon
NGA	National Geospatial Agency
OSI	Open Systems Interconnection
OUSD-R&E	Office of the Under Secretary of Defense for Research and Engineering
PC	personal computer
PLRACTA	Position Location Reporting and Control of Tactical Aircraft
RFC	request for comments
SAR	synthetic aperture radar
SIPRNet	Secret Internet Protocol Router Network
SoS	system of systems
STANAG	Standardization Agreement
STARS	Surveillance Target Attack Radar System
TADIL J	Tactical Digital Information Link J
TCP/IP	transmission control protocol/internet protocol
TDL	tactical data link
TDMA	time-division multiple access
TEL	transporter erector launcher

TST	technical support team
UAS	unmanned aerial system
UC2	universal command and control
VCR	videocassette recorder
XML	eXtensible Markup Language

References

AEDP-7—*See* Allied Engineering Documentation Publication Number 7.

Alberts, David S., and Richard E. Hayes, *Power to the Edge: Command . . . Control . . . in the Information Age*, Washington, D.C.: U.S. Department of Defense, Command and Control Research Program, 2003.

Allen, Robert H., and Ram D. Sriram, "The Role of Standards in Innovation," *Technological Forecasting and Social Change*, Vol. 64, No. 2–3, 2000, pp. 171–181.

Allied Engineering Documentation Publication Number 7, *NATO Ground Moving Target Indicator (GMTI) Format, STANAG 4607 Implementation Guide*, Brussels, Belgium: NATO Standardization Agency, May 2013.

Army Training and Doctrine Command, *TADIL J: Introduction to Tactical Digital Information Link J and Quick Reference Guide*, Fort Monroe, Va.: Defense Technical Information Center, 2000.

Beaulne, Pierre D., *The Addition of Enhanced Capabilities to NATO GMTIF STANAG 4607 to Support RADARSAT-2 GMTI Data*, Ottawa, Canada: Defense Research and Development, technical memorandum DRDC-O-TM-2007-341, December 2007.

Blasko, Dennis J., "'Technology Determines Tactics': The Relationship Between Technology and Doctrine in Chinese Military Thinking," *Journal of Strategic Studies*, Vol. 34, No. 3, 2011, pp. 355–381.

Cargill, Carl F., "Why Standardization Efforts Fail," *Standards*, Vol. 14, No. 1, Summer 2011.

Copp, Tara, "Space Force Aims to Take on an Air Force Surveillance Mission," *Defense One*, May 12, 2021. As of July 9, 2021:
https://www.defenseone.com/technology/2021/05/space-force-aims-take-air-force-surveillance-mission/173997/

Coté, Owen R., *The Politics of Innovative Military Doctrine: The U.S. Navy and Fleet Ballistic Missiles*, dissertation, Boston, Mass.: Massachusetts Institute of Technology, 1996. As of July 7, 2021:
http://edocs.nps.edu/AR/topic/theses/1996/Feb/96Feb_Cote_PhD.pdf

David, Paul A., "Clio and the Economics of QWERTY," *American Economic Review*, Vol. 75, No. 2, May 1985, pp. 332–337.

Davis, Barry W., Cecil Graham, David Stamm, and Chris Parker, "Tactical Digital Information Link (TADIL) J Range Extension (JRE)," *MILCOM 97 Proceedings,* Vol. 1, Monterey, Calif.: IEEE and AFCEA, November 1997, pp. 408–412.

Dimarogonas, James, Jasmin Leveille, Jan Osburg, Shane Tierney, Bonnie Triezenberg, Graham Andrews, Bryce Downing, Muharrem Mane, and Monica Rico, *Universal Command and Control Language Early Engineering Study, Performance Effects of a Universal Command and Control Standard*, Santa Monica, Calif.: RAND Corporation, 2021, Not available to the general public.

Dorris, Steven A., David E. Corman, Thomas S. Herm, and Eric Martens, "TCP/IP Tunneling Protocol for Link 16," U.S. Patent 7953110B1, filed January 11, 2010, assigned May 31, 2011. As of July 12, 2021:
https://patents.google.com/patent/US7953110B1/en

Dranove, David, and Neil Gandal, "The Dvd-vs.-Divx Standard War: Empirical Evidence of Network Effects and Preannouncement Effects," *Journal of Economics & Management Strategy*, Vol. 12, No. 3, 2003, pp. 363–386.

Drezner, Jeffrey A., Jon Schmid, Justin Grana, Megan McKernan, and Mark Ashby, *Benchmarking Data Use and Analytics in Large, Complex Private-Sector Organizations: Implications for Department of Defense Acquisition*, Santa Monica, Calif.: RAND Corporation, RR-A225-1, 2020. As of October 21, 2021:
https://www.rand.org/pubs/research_reports/RRA225-1.html

Dyer, Davis, and Michael A. Dennis, *Architects of Information Advantage: The MITRE Corporation Since 1958*, McLean, Va.: Community Communications Corp., 1998.

Egyedi, T. M., "Institutional Dilemma in ICT Standardization: Coordinating the Diffusion of Technology," in Kai Jakobs, ed., *Information Technology Standards and Standardization: A Global Perspective*, Hershey, Pa.: IGI Global, 2000, pp. 48–62.

Elledge, Ashley, Terrance Zimmerman, Bruce Robinson, and Jasen Jacobsen, "Exchanging Link 16 CoT Messages and US Army ABCS Messages via PASS," *MILCOM 2007, IEEE Military Communications Conference*, Orlando, Fla., October 2007, pp. 1–5.

Farrell, Joseph, and Paul Klemperer, "Chapter 31: Coordination and Lock-In: Competition with Switching Costs and Network Effects," in M. Armstrong and R. Porter, eds., *Handbook of Industrial Organization*, Vol. 3, Amsterdam: North Holland Publishing, 2007, pp. 1967–2072.

Frey, Steven E., *A Strategic Framework for Program Managers to Improve Command and Control System Interoperability*, thesis, Boston, Mass.: Massachusetts Institute of Technology, 2002.

Gonzales, Daniel, John S. Hollywood, Gina Kingston, and David Signori, *Network-Centric Operations Case Study: Air-to-Air Combat With and Without Link 16*, Santa Monica, Calif.: RAND Corporation, MG-268-OSD, 2005. As of July 12, 2021:
https://www.rand.org/pubs/monographs/MG268.html

Gonzales, Daniel, Daniel M. Norton, and Myron Hura, *Multifunctional Information Distribution System (MIDS) Program Case Study*, Santa Monica, Calif.: RAND Corporation, DB-292-AF, 2000. As of July 12, 2021:
https://www.rand.org/pubs/documented_briefings/DB292.html

Griffin, Michael D., "Statement of Michael D. Griffin, Under Secretary of Defense for Research and Engineering," statement before the House Armed Services Subcommittee on Intelligence, Emerging Threats and Capabilities, FY2020 Science and Technology Posture Hearing, Washington, D.C., March 11, 2020.

Grindley, Peter, *Standards, Strategy, and Policy: Cases and Stories*, New York: Oxford University Press, 1995.

Hanseth, Ole, Eric Monteiro, and Morten Hatling, "Developing Information Infrastructure: The Tension Between Standardization and Flexibility," *Science, Technology, and Human Values*, Vol. 21, No. 4, Fall 1996, pp. 407–426.

Heaney, Matthew, "Why Ada Isn't Popular," AdaPower, December 8, 1998.

Hicks, Kathleen, "Creating Data Advantage," memorandum from the Deputy Secretary of Defense to Senior Pentagon Leadership Commanders of the Combatant Commands Defense Agency and U.S. Department of Defense Field Activity Directors, Washington, D.C., May 5, 2021. As of July 7, 2021:
https://media.defense.gov/2021/May/10/2002638551/-1/-1/0/
DEPUTY-SECRETARY-OF-DEFENSE-MEMORANDUM.PDF

Ho, Jae-Yun, and Eoin O'Sullivan, "Dimensions of Standards for Technological Innovation—Literature Review to Develop a Framework for Anticipating Standardisation Needs," *EURAS Proceedings 2015: The Role of Standards in Transatlantic Trade and Regulation*, 2015.

Hoffman, Paul, and Susan Harris, *The Tao of the IETF: A Novice's Guide to the Internet Engineering Task Force*, RFC 4677, September 2006. As of July 8, 2021:
https://datatracker.ietf.org/doc/html/rfc4677

Huckins, Clem H., *Evolution of a Standard: The STANAG 4607 NATO GMTI Format*, McLean, Va.: MITRE Corporation, technical paper, February 2005. As of October 21, 2021:
https://www.mitre.org/sites/default/files/pdf/05_0164.pdf

Hura, Myron, Gary McLeod, Eric V. Larson, James Schneider, Daniel Gonzales, Daniel M. Norton, Jody Jacobs, Kevin M. O'Connell, William Little, Richard Mesic, and Lewis Jamison, *Interoperability: A Continuing Challenge in Coalition Air Operations*, Santa Monica, Calif.: RAND Corporation, MR-1235-AF, 2000. As of May 10, 2021:
https://www.rand.org/pubs/monograph_reports/MR1235.html

International Command and Control Institute, homepage, undated. As of July 8, 2021:
http://internationalc2institute.org/

International Organization for Standardization and the International Electrotechnical Commission, "Information Technology—Open Systems Interconnection: Basic Reference Model: The Basic Model—Part 1," ISO/IEC 7498-1:1994, webpage, 1994. As of July 8, 2021:
https://www.iso.org/obp/ui/#iso:std:iso-iec:7498:-1:ed-1:v2:en

ISO and the International Electrotechnical Commission—*See* International Organization for Standardization and the International Electrotechnical Commission.

Jassemi-Zargani, Rahim, and Pierre D. Beaulne, "SBR MTI Impact on Coastal and Border Surveillance," meeting proceedings, NATO Standardization Agency, Science and Technology Organizations, RTO-MP-SET-125-21, April 18, 2008, pp. 21-1–21-16.

"Joint Tactical Information Distribution System," *Wikipedia*, last updated October 9, 2020. As of July 12, 2021: https://en.wikipedia.org/w/index.php?title=Joint_Tactical_Information_Distribution_System&oldid=982715392

Kreitmair, Thomas, Joe Ross, and Trond Skaar, *Experimentation Activities with Aerospace Ground Surveillance*, The Hague, Netherlands: NATO Consultation Command and Control Agency, AD-A460950; X5-NATO/C3/NL, June 2005.

Larson, Eric V., Gustav Lindstrom, Myron Hura, Ken Gardiner, Jim Keffer, and William Little, *Interoperability of U.S. and NATO Allied Air Forces: Supporting Data and Case Studies*, Santa Monica, Calif.: RAND Corporation, MR-1603-AF, 2004. As of July 12, 2021: https://www.rand.org/pubs/monograph_reports/MR1603.html

Levy, Jack S., "Case Studies: Types, Designs, and Logics of Inference," *Conflict Management and Peace Science*, Vol. 25, No. 1, 2008, pp. 1–18.

Liebowitz, Stan J., and Stephen E. Margolis, "The Fable of the Keys," *Journal of Law and Economics*, Vol. 33, No. 1, April 1990, pp. 1–25.

———, "Network Externality: An Uncommon Tragedy," *Journal of Economic Perspectives*, Vol. 8, No. 2, 1994, pp. 133–150.

Marques, Mário Monteiro, *STANAG 4586—Standard Interfaces of UAV Control System (UCS) for NATO UAV Interoperability*, Portugal: NATO Standardization Agency, Science and Technology Organizations, STO-EN-SCI-271, May 1, 2015.

Mattis, James, *Summary of the 2018 National Defense Strategy of the United States of America*, Washington, D.C.: U.S. Department of Defense, 2018.

Missile Defense Advocacy Alliance, "Link 16," webpage, January 2017. As of July 12, 2021: https://missiledefenseadvocacy.org/defense-systems/link-16/

Northrop Grumman, *Understanding Voice and Data Link Networking: Northrop Grumman's Guide to Secure Tactical Data Links*, San Diego, Calif., No. 135-02-005, December 2014.

Office of the Secretary of the Air Force for Acquisition, *JTIDS Operational Special Project (OSP) Report to Congress, Mission Area Director for Information Dominance*, Washington, D.C.: Headquarters U.S. Air Force, December 1997.

Office of the Secretary of Defense, *Unmanned Aerial Vehicles Roadmap: 2002–2027*, Washington, D.C.: U.S. Department of Defense, 2002.

O'Hern, Brian R., Custodian, NATO STANAG 4607 (GMTI), Air Force Research Laboratory (AFRL/RIEA), "STANAG 4607 Interoperability Testing," memorandum to David Schneeberger, Chairman, Joint Capability Group for ISR (JCGISR), Rome, N.Y., March 20, 2019. As of July 8, 2021: https://diweb.hq.nato.int/nafag2/jcgisr/STANAGs/STANAG%204607%20Test%20Agent%20assignment%2020Mar19.pdf

Posen, Barry R., *The Sources of Military Doctrine: France, Britain, and Germany Between the World Wars*, Cornell Studies in Security Affairs, Ithaca, N.Y.: Cornell University Press, 1986.

Rose, Marshall. T., "The Future of OSI: A Modest Prediction," *ULPAA '92: Proceedings of the IFIP TC6/WG6.5 International Conference: Upper Layer Protocols, Architectures and Applications*, Vol. C-7, 1992, pp. 367–376.

Rosen, Stephen P., *Winning the Next War: Innovation and the Modern Military*, Ithaca, N.Y.: Cornell University Press, 1994.

Sapolsky, Harvey M., *The Polaris System Development: Bureaucratic and Programmatic Success in Government*, Cambridge, Mass.: Harvard University Press, 1972.

Schmid, Jon, *The Determinants of Military Technology Innovation and Diffusion*, dissertation, Atlanta, Ga.: Georgia Institute of Technology, 2018a. As of July 7, 2021: https://smartech.gatech.edu/bitstream/handle/1853/59877/SCHMID-DISSERTATION-2018.pdf

———, "The Diffusion of Military Technology," *Defence and Peace Economics*, Vol. 29, No. 6, 2018b, pp. 595–613.

——, "Intelligence Innovation: Sputnik, the Soviet Threat, and Innovation in the US Intelligence Community," in Margaret E. Kosal, ed., *Technology and the Intelligence Community*, New York: Springer International Publishing, 2018c, pp. 39–53.

Schmid, Jon, and Jonathan Huang, "State Adoption of Transformative Technology: Early Railroad Adoption in China and Japan," *International Studies Quarterly*, Vol. 61, No. 3, September 2017, pp. 570–583.

Schmidt, Charles M., *Best Practices for Technical Standard Creation*, McLean, Va.: MITRE Corporation, technical paper, 2017. As of July 7, 2021:
https://www.mitre.org/publications/technical-papers/best-practices-for-technical-standard-creation

Schuler, Douglas, and Aki Namioka, eds., *Participatory Design: Principles and Practices*, Boca Raton, Fla.: CRC Press, 1993.

Shin, Dong-Hee, Hongbum Kim, and Junseok Hwang, "Standardization Revisited: A Critical Literature Review on Standards and Innovation," *Computer Standards and Interfaces*, Vol. 38, February 2015, pp. 152–157.

STANAG 4607—*See* Standardization Agreement 4607.

Standardization Agreement 4607, *NATO Ground Moving Target Indicator (GMTI) Format*, Brussels, Belgium: North Atlantic Treaty Organization Standardization Agency, September 2010.

Swedberg, Richard, "Exploratory Research," in Colin Elman, John Gerring, and James Mahoney, eds., *The Production of Knowledge: Enhancing Progress in Social Science*, Cambridge, United Kingdom: Cambridge University Press, 2020, pp. 17–41.

U.S. Department of Defense, *C3 Command, Control, and Communications: Modernization Strategy*, Washington, D.C., September 2020.

Wang, Jaesun, and Seoyong Kim, "Time to Get In: The Contrasting Stories About Government Interventions in Information Technology Standards (the Case of CDMA and IMT-2000 in Korea)," *Government Information Quarterly*, Vol. 24, No. 1, January 2007, pp. 115–134.

White, B. E., "Tactical Data Links, Air Traffic Management, and Software Programmable Radios," *Gateway to the New Millennium, 18th Digital Avionics Systems Conference Proceedings*, Vol. 1, St. Louis, Mo., October 1999, pp. 5.C.5-1–5.C.5-8.

Williams, Lawrence, "IPv4 vs IPv6: What's the Difference Between IPv4 and IPv6?" Guru99, webpage, last updated October 7, 2021. As of October 21, 2021:
https://www.guru99.com/difference-ipv4-vs-ipv6.html#:~:text=KEY%20DIFFERENCE,separated%20by%20a%20colon(%3A)A

Wilson, Warren J., "Applying Layering Principles to Legacy Systems: Link 16 as a Case Study," *2001 MILCOM Proceedings Communications for Network-Centric Operations: Creating the Information Force*, Vol. 1, McLean, Va., October 2001, pp. 526–531.

Yin, Robert K., *Case Study Research and Applications: Design and Methods*, 6th ed., Thousand Oaks, Calif.: Sage Publications, Inc., October 2017.